A PRIMER OF
NONLINEAR ANALYSIS

D0075215

A Primer of
Nonlinear Analysis

Antonio Ambrosetti
Scuola Normale Superiore, Pisa

Giovanni Prodi
Department of Mathematics, University of Pisa

CAMBRIDGE
UNIVERSITY PRESS

Published by the Press Syndicate of the University of Cambridge
The Pitt Building, Trumpington Street, Cambridge CB2 1RP
40 West 20th Street, New York, NY 10011-4211, USA
10 Stamford Road, Oakleigh, Victoria 3166, Australia

© Cambridge University Press 1993

First published 1993

Printed in Great Britain at the University Press, Cambridge

Library of Congress cataloguing in publication data available
A catalogue record for this book is available from the British Library

ISBN 0 521 37390 5   hardback

Contents

Preface

In the last few decades, once linear functional analysis was quite widely and thoroughly estabilished, the interest of scientists in Nonlinear Analysis has been increasing a lot. On the one hand the treatments of various classical problems have been unified; on the other, theories specifically nonlinear, of great significance and applicability, have come out.

This book provides an introduction to basic aspects of Nonlinear Analysis, namely those based on differential calculus in Banach spaces. The matter is expressed in a geometric style, in the sense that the results obtained are often a transposition to infinite dimensions of events which are intuitive in $\mathbb{R}^2$ or $\mathbb{R}^3$. Indeed, this was a primary characteristic of the works of Pincherle, Volterra and Fréchet.

The topics treated can be divided into two main parts and are preceded by a short chapter in which some introductory material is recalled, and also the main notation fixed.

In the first part, differential calculus in Banach spaces is discussed, together with local and global inversion theorems.

The second part deals with bifurcation theory which in spite of its elementary character is, perhaps, one of the most powerful tools used in Nonlinear Analysis. Our attention is here devoted almost entirely to the case of simple eigenvalues, but an accurate analysis of hypotheses is made, in order to include, for example, also the celebrated Hopf theorem.

A specific feature of Nonlinear Analysis is that the theoretical setting is strictly linked to applications, especially those related to differential equations, where the power of nonlinear methods is expressed in a more striking way. Moreover, a relevant fact to be emphasized is that problems that are often considered of formidable difficulty, once they are framed in an appropriate functional setting, may be faced and solved quite easily.

It is, indeed, this aspect, peculiar to Nonlinear Analysis, that has driven us to leave considerable space to applications to differential equations, including various important classical problems such as Bénard Problem, the problem of water waves, the restricted three-body problem and some others. Thus, in addition to more elementary examples and applications that usefully accomplish theoretical results, in separate paragraphs and/or chapters, we deal with those problems which require more care both in formulation and in resolution.

Tools, still of remarkable importance, such as the theory of Leray–Schauder topological degree, or the critical point theory, which would require wider theoretical background and more subtle arguments, are left out in this treatise.

The book in its outlines is self-contained for a reader who, besides infinitesimal calculus, is acquainted with fundamental results of Linear Functional Analysis such as the Hahn–Banach Theorem, the "Closed Graph" Theorem and the Fredholm Alternative Principle. Only some of the problems dealing with partial differential equations require a certain knowledge of Sobolev spaces and therefore, in just a few cases, we refer to results contained in original papers.

This volume is partially based on an earlier booklet, published in Italian by the Scuola Normale Superiore di Pisa in the series "Quaderni". The authors wish to thank the Scuola Normale Superiore for the encouragement.

0

Preliminaries and notation

This chapter contains the notation and some preliminary tools used throughout the book. Almost always, the results are not quoted in the most general form, but in a way appropriate to our purposes; nevertheless some of them are actually slightly more general than we strictly need. For more detail we refer to any book of (linear) Functional Analysis (for example [Y] or [Br] for topics reviewed in sections 0.1–0.4; to [Br], [KFS], [GT] for 0.5–0.6).

0.1   Some notation and definitions

$\mathbb{R}^n$ will denote the n-dimensional Euclidian space with scalar product $x \cdot y$ and norm given by $|x|^2 = x \cdot x$.

$X, Y, Z, \ldots$ denote (real) Banach spaces with norm $\|.\|_X$, $\|.\|_Y$, etc., respectively (the subscript will be omitted if no possible confusion arises). $B(x^*, r)$ denotes the ball $\{x \in X : \|x - x^*\| < r\}$ and $B(r)$ stands for $B(0, r)$.

If X^* is the topological dual of X the symbol $\langle ., . \rangle$ will indicate the duality pairing between X and X^*.

Let $\{x_n\}$ be a sequence in X. We say that x_n converges (strongly) to $x \in X$, written as $x_n \to x$, if $\|x_n - x\| \to 0$ as $n \to \infty$; we say that x_n converges weakly to x, written as $x_n \rightharpoonup x$, if $\langle \psi, x_n - x \rangle \to 0$ as $n \to \infty$ for all $\psi \in X^*$.

Let X be a Banach space and let V be a closed subspace of X. A *topological complement* of V in X is a *closed* subspace W of X such that $V \cap W = \{0\}$ and $X = V \oplus W$; $V \oplus W$ is called a *splitting* of X.

Recall also that, associated with such a splitting, there are (continuous) projections P and Q onto V and W respectively.

0.2 Continuous mappings

We will deal with continuous maps $f : U \rightarrow Y$, where U is an open subset of X. Continuity means that $f(x_n) \rightarrow f(x)$ (strongly) for any sequence x_n strongly convergent to $x \in X$. The set of all continuous $f : U \rightarrow Y$ will be denoted by $C(U, Y)$.

0.3 Integration

For continuous maps $f : [a, b] \rightarrow Y$ the definition of the Cauchy integral is given as in the elementary case, as the (strong) limit of the finite sums $\Sigma f(\xi_i)(t_i - t_{i-1})$ (with obvious meaning).

From

$$\|\Sigma_i f(\xi_i)(t_i - t_{i-1})\| \leq \Sigma_i \|f(\xi_i)(t_i - t_{i-1})\| \leq \Sigma_i \|f(\xi_i)\|(t_i - t_{i-1})$$

there follows immediately the inequality

$$\left\|\int\limits_a^b f(t)\mathrm{d}t\right\| \leq \int\limits_a^b \|f(t)\|\mathrm{d}t.$$

0.4 Linear continuous maps

The space of linear continuous maps $A : X \rightarrow Y$ will be denoted by $L(X, Y)$. The range of $A, R(A)$, is the linear space $\{A(x) : x \in X\}$. Sometimes, when $Y = X$, we will use the notation $L(X)$ instead of $L(X, X)$. Equipped with the norm

$$\|A\| = \sup\{\|A(x)\| : \|x\| \leq 1\},$$

$L(X, Y)$ is a Banach space. The identity map in $L(X)$ will be denoted by I_X.

Hereafter, for linear maps, the notation Ax or $A[x]$ may replace $A(x)$.

An *eigenvalue* of $A \in L(X)$ is a $\mu \in \mathbb{C}$ such that the equation $Ax = \mu x$ has solutions $x \neq 0$. Any such solution is an *eigenvector* associated to μ and Ker $(\mu I - A)$ is the *eigenspace* associated to μ. We will be mainly interested in the case when $A \in L(X)$ is compact, namely when A is completely continuous, if this is the case, the following results hold.

Theorem 0.1 (Fredholm Alternative) *Let $A \in L(X)$ be compact and $\mu \neq 0$. Then*

(i)   Ker $(\mu I - A) = \{0\}$ *if and only if $R(\mu I - A) = X$,*

(ii) $R(\mu I - A) = [\text{Ker } (\mu I - A^*)]^\perp = \{u \in X : \langle \psi, u \rangle = 0 \text{ for all } \psi \in$
       $\text{Ker } (\mu I - A^*)\}.$

Moreover one has the following

Theorem 0.2   *Let* $A \in L(X)$ *be compact and* $\mu \neq 0$. *Then*

(i) $\text{Ker } (\mu I - A)$ *is finite-dimensional and* $\text{Range}(\mu I - A)$ *is closed,*

(ii)   *the sequence* $\text{Ker } ((\mu I - A)^n)$ $(n \in \mathbb{N})$ *is increasing, that is* Ker
       $((\mu I - A)^m) \subset \text{Ker } ((\mu I - A)^{m+1})$ *for all* $m \geq 1$,

(iii) *there exists a finite* $p \in \mathbb{N}$ *such that* $\text{Ker } ((\mu I - A)^p) = \text{Ker } ((\mu I - A)^q)$ *if and only if* $q \geq p$.

The (algebraic) multiplicity of μ is the dimension of the linear subspace

$$\cup_{n \in \mathbb{N}} \text{Ker}((\mu I - A)^n) = \text{Ker}((\mu I - A)^p).$$

It is worth pointing out that the *algebraic* multiplicity of μ is, in general, different from the *geometric* multiplicity, defined as the dimension of $\text{Ker } (\mu I - A)$ (algebraic and geometric multiplicity coincide for self-adjoint operators on Hilbert spaces). Hereafter, by the multiplicity of an eigenvalue $\mu \neq 0$ of a completely continuous $A \in L(X)$ we will always mean the algebraic multiplicity. An eigenvalue will be said to be *simple* if its multiplicity is 1.

0.5   Function spaces

Let Ω be an open subset of $\mathbb{R}^n$ with boundary $\partial\Omega$ and closure $\overline{\Omega}$.
We will use standard notation for spaces of continuous or differentiable real-valued functions $C^k(\overline{\Omega})$ $(k \geq 0)$, for Lebesgue spaces $L^p(\Omega)$ $(1 \leq p < \infty)$ or $L^\infty(\Omega)$. In some cases we will write $C(\overline{\Omega})$ instead of $C^0(\overline{\Omega})$. The spaces above are Banach spaces under the norms defined, respectively, by

$$\|u\|_C = \sup \{|u(x)| : x \in \overline{\Omega}\},$$

$$\|u\|_{C^k} = \sum_{0 \leq |\beta| \leq k} \|D^\beta u\|_C \quad (\beta \text{ is a multi index}),$$

$$\|u\|_{L^p} = \left[\int_\Omega |u|^p \right]^{1/p}$$

(the symbol $\mathrm{d}x$ will be omitted whenever there is no ambiguity)

$$\|u\|_{L^\infty} = \text{ess sup } \{|u(x)| : x \in \Omega\}.$$

For $k \geq 0$ and $0 < \alpha \leq 1$, $C^{k,\alpha}(\overline{\Omega})$ denotes the space of Hölder functions with exponent α, namely the $u \in C^k(\overline{\Omega})$ such that, for all multi-index $\beta, |\beta| = k$,

$$\sup \left\{ \frac{|D^\beta u(x) - D^\beta u(y)|}{|x - y|^\alpha} : x, y \in \Omega, x \neq y \right\} < \infty.$$

For $k = 0$ and $\alpha = 1$, $C^{0,1}(\overline{\Omega})$ is nothing but the space of Lipschitz-continuous functions on $\overline{\Omega}$.

Equipped with the norm

$$\|u\|_{C^{k,\alpha}} = \|u\|_{C^k} +$$

$$+ \sup \left\{ \frac{|D^\beta u(x) - D^\beta u(y)|}{|x - y|^\alpha} : x, y, \in \Omega, x \neq y, \, |\beta| = k \right\},$$

$C^{k,\alpha}(\overline{\Omega})$ is a Banach space.

In some cases, we shall also work with Sobolev spaces $H^{k,p}(\Omega)$ ($k \geq 1, p \in [1, \infty)$) equipped with the norm

$$\|u\|_{H^{k,p}} = \sum_{0 \leq |\beta| \leq k} \|D^\beta u\|_{L^p}.$$

The notation H^k will stand for $H^{k,2}$ while $H_0^k(\Omega)$ will denote the closure of $C_0^\infty(\Omega)$, the space of C^∞ functions with compact support in Ω, under the norm $\|u\|_{H^{k,2}}$. Among others, let us recall the following result.

Theorem 0.3   (Poincaré Inequality) *Let Ω be bounded. Then there exists a constant $c = c(\Omega)$ such that*

$$\int_\Omega |u|^2 \leq c \int_\Omega |\nabla u|^2 \text{ for all } u \in H_0^1(\Omega).$$

As a consequence, $\|\nabla u\|_{L^2}$ is a norm in $H_0^1(\Omega)$ equivalent to $\|u\|_{H^{1,2}}$.

In addition to the Poincaré Inequality one has that the *embedding* of $H_0^1(\Omega)$ in $L^2(\Omega)$ is compact (Rellich's Theorem). Let us recall that X is embedded in Y, $X \hookrightarrow Y$, if $X \subset Y$ and the inclusion $\imath : X \to Y$ is continuous. If $X \hookrightarrow Y$ then $\exists \, c > 0$ such that

$$\|u\|_Y \leq c\|u\|_X, \text{ for all } u \in X.$$

If the inclusion $\imath : X \to Y$ is *compact* we will write $X \hookrightarrow\hookrightarrow Y$.

The following result is a particular case of the "Sobolev Embedding Theorems".

**Theorem 0.4   ** *Suppose that Ω is bounded open set in $\mathbb{R}^n$, with boundary $\partial\Omega$ of class $C^{0,1}$, and let $k \geq 1$ and $1 \leq p \leq \infty$.*

(i)   *If $kp < n$, then $H^{k,p}(\Omega) \hookrightarrow L^q(\Omega)$, for all $1 \leq q \leq np/(n - kp)$.*

(ii)   *If $kp = n$, then $H^{k,p}(\Omega) \hookrightarrow L^q(\Omega)$, for all $q \in [1, \infty)$.*

(iii) *If $kp > n$, then $H^{k,p}(\Omega) \hookrightarrow C^{0,\alpha}(\overline{\Omega})$, where $\alpha = k-n/p$ if $k-n/p < 1$; $\alpha \in [0,1)$ is arbitrary if $k - n/p = 1$ and $p > 1$; $\alpha = 1$ if $k - n/p > 1$.*

In addition, there result the following.

(i′) *If $kp < n$, then $H^{k,p}(\Omega) \hookrightarrow\hookrightarrow L^q(\Omega)$, for all $1 \le q < np/(n - kp)$.*
(ii′) *If $kp = n$, then $H^{k,p}(\Omega) \hookrightarrow\hookrightarrow L^q(\Omega)$, for all $q \in [1,\infty)$.*
(iii′) *If $kp > n$, then $H^{k,p}(\Omega) \hookrightarrow\hookrightarrow C(\overline{\Omega})$.*

0.6 Elliptic boundary value problems

Let Ω be a bounded domain (i.e. open connected) in $\mathbb{R}^n$ with smooth boundary $\partial\Omega$ (this will always be understood hereafter) and let $\mathcal{L}$ denote the differential operator

$$\mathcal{L} = \sum_{1\le i,j\le n} \frac{\partial}{\partial x_i}\left(a_{ij}(x)\frac{\partial}{\partial x_j}\right) \tag{0.1}$$

where

$$a_{ij} = a_{ji} \in C^\infty(\overline{\Omega}). \tag{0.2}$$

$\mathcal{L}$ is (uniformly) elliptic if there exists $\alpha > 0$ such that

$$\sum_{1\le i,j\le n} a_{ij}(x)\xi_i\xi_j \ge \alpha|\xi|^2, \text{ for all } x \in \Omega \text{ and } \xi \in \mathbb{R}^n \tag{0.3}$$

Throughout the book, any elliptic operator will be an *elliptic operator with smooth coefficients*, namely an $\mathcal{L}$ of the form (0.1) and such that (0.2)–(0.3) hold.

Consider the Dirichlet Boundary Value Problem (b.v.p. for short)

$$\left.\begin{array}{l} -\mathcal{L}u = h(x) \text{ in } \Omega, \\ u = 0 \text{ on } \partial\Omega, \end{array}\right\} \tag{0.4}$$

where h is given a function on Ω.

Let $h \in L^2(\Omega)$; a *weak solution* of (0.4) is a $u \in H_0^1(\Omega)$ such that

$$\sum_{1\le i,j\le n} \int_\Omega a_{ij}\frac{\partial u}{\partial x_i}\frac{\partial v}{\partial x_j} = \int_\Omega hv, \text{ for all } v \in C_0^\infty(\Omega).$$

If u is a weak solution of (0.4) and $u \in C^2(\Omega)$, then u is a classical solution.

Theorem 0.5 *Suppose $\mathcal{L}$ is an elliptic operator. Then the following results hold.*

(i) *Let $h \in L^p(\Omega)$, $2 \le p < \infty$. Then (0.4) has a unique (weak) solution $u \in H_0^1(\Omega) \cap H^{2,p}(\Omega)$ and the following estimate holds:*

$$\|u\|_{H^{2,p}} \le c\|h\|_{L^p}.$$

(ii)   *If $h \in L^\infty(\Omega)$ then $u \in C^{1,\alpha}(\overline{\Omega})$ for any $0 < \alpha < 1$ and*

$$\|u\|_{C^{1,\alpha}} \leq c \, \|h\|_{L^\infty}.$$

(iii) *If $h \in C^{0,\alpha}(\overline{\Omega})$ then $u \in C^{2,\alpha}(\overline{\Omega})$ is a classical solution of (0.4) and*

$$\|u\|_{C^{2,\alpha}} \leq c \, \|h\|_{C^{0,\alpha}}.$$

In the above c stands for a positive constant, depending on Ω.

As a consequence of the preceding results, we can define an operator $K : L^2(\Omega) \to L^2(\Omega)$ (the *Green* operator of $-\mathcal{L}$ with zero Dirichlet boundary conditions) setting $Ku = v$ if and only if $-\mathcal{L}v = u$, $v \in H_0^1(\Omega)$. From the Rellich Theorem it follows immediately that K is compact.

Given a function $m \in L^\infty(\Omega)$, let us consider the linear eigenvalue problem

$$\left.\begin{array}{r} -\mathcal{L}u = \lambda m u \text{ in } \Omega, \\ u = 0 \text{ on } \partial\Omega. \end{array}\right\} \tag{0.5}$$

An eigenvalue of (0.5) is a λ such that (0.5) has a solution $u \neq 0$. Any $\phi \neq 0$ satisfying (0.5) is an eigenfunction associated to the eigenvalue λ. If we set $\mu = 1/\lambda$ and $K_m(u) = K(mu)$, problem (0.5) is equivalent to $\mu u = K_m u$. The eigenvalues λ_k of (0.5) correspond, through $\mu_k = 1/\lambda_k$ to the eigenvalues of K_m. The multiplicity of λ_k is the multiplicity of μ_k. In some cases we will write $\lambda_k(m)$ or $\lambda_k(\Omega)$ to highlight the dependence of the eigenvalues of (0.5) on m or Ω.

Theorem 0.6 *Let $m \in L^\infty(\Omega)$, $m \geq 0$ and $m(x) > 0$ in a set of positive measure.*

(i)   *Problem (0.5) has a sequence*

$$0 < \lambda_1(m) < \lambda_2(m) \leq \ldots \leq \lambda_k(m) \leq \ldots$$

of eigenvalues such that $\lambda_k(m) \to +\infty$ as $k \to \infty$. The first eigenvalue $\lambda_1(m)$ is simple and the corresponding eigenfunctions do not change sign in Ω. We will let denote ϕ_1, (sometimes only ϕ) the eigenfunction such that (a) $\phi > 0$ in Ω and (b) $\int_\Omega \phi^2 = 1$.
We will also let ϕ_k denote the eigenfunctions corresponding to λ_k normalized by

$$\int_\Omega \phi_h \phi_k = \delta_{hk} = \begin{cases} 1 & \text{if } h = k, \\ 0 & \text{if } h \neq k. \end{cases}$$

When $m \equiv 1$ we will simply write λ_k instead of $\lambda_k(1)$.

(ii)   *(Comparison property) If $m \leq M$ in Ω then $\lambda_k(m) \geq \lambda_k(M)$; if $m < M$ in a subset of positive measure then $\lambda_k(m) > \lambda_k(M)$. In particular, if $m < \lambda_k$ (resp.$> \lambda_k$) then $\lambda_k(m) > 1$ (resp.< 1).*

(iii) (Variational characterization) *There results*

$$\lambda_k(m) = \max \left\{ \int_\Omega mv^2 : v \in H_0^1(\Omega), \quad \int_\Omega \sum a_{ij} \frac{\partial u}{\partial x_i} \frac{\partial u}{\partial x_j} = 1 \, , \right.$$

$$\left. \int_\Omega v\phi_i = 0, \text{ for all } i = 1, \ldots, k-1 \right\}.$$

(iv) (Continuity property) $\lambda_1(m)$ *depends continuously on m in the* $L^{n/2}(\Omega)$ *topology.*

(v) *Let Ω' be a bounded domain, such that $\Omega' \subset \Omega$. Then $\lambda_k(\Omega') \geq \lambda_k(\Omega)$ for all $k \geq 1$.*

Consider the non-homogenous b.v.p.

$$\left. \begin{array}{c} -\mathcal{L}u = \lambda mu + h \text{ in } \Omega, \\ u = 0 \text{ on } \partial\Omega, \end{array} \right\} \tag{0.6}$$

with, say, $h \in L^2(\Omega)$.

From the Fredholm Alternative Theorem 0.1 we get the following.

Theorem 0.7

(i) *If λ is not an eigenvalue of (0.5), then (0.6) has a unique solution for all $h \in L^2(\Omega)$;*

(ii) *if λ is an eigenvalue of (0.5), then (0.6) has a solution if and only if $\int_\Omega h\phi_k = 0$ for any k such that $\lambda = \lambda_k$.*

According to Theorem 0.4 (iii) all the preceding discussion can be carried over taking $X = C^{2,\alpha}(\overline{\Omega})$, $h \in C^{0,\alpha}(\overline{\Omega})$ and m smooth.

The arguments above apply to Sturm–Liouville Problems

$$\left. \begin{array}{c} -\dfrac{\mathrm{d}}{\mathrm{d}x}\left(\alpha \dfrac{\mathrm{d}}{\mathrm{d}x}u\right) + \beta u = h(x) \quad (0 < x < \pi), \\ a_0 u(0) + b_0 u'(0) = a_1 u(\pi) + b_1 u'(\pi) = 0, \end{array} \right\}$$

where $\alpha \in C^1([0,\pi])$, $\beta \in C((0,\pi])$, $\alpha, \beta > 0$ on $[0,\pi]$, and a_0, b_0, a_1, b_1 are such that $(a_0^2 + b_0^2)(a_1^2 + b_1^2) \neq 0$.

In fact, it is known [**D1**] that for all $h \in X := C([0,\pi])$ there exists a unique $u \in C^2([0,\pi])$ satisfying (0.6) and hence the map $K : h \to K(h)$ (is linear and) as an operator from X into itself is compact. It is also known that such a K has a sequence of positive, *simple* eigenvalues $\mu_1 > \mu_2 > \ldots > \mu_k \ldots$, such that $\mu_k \to 0$ as $k \to \infty$. Correspondingly, the linear Sturm–Liouville eigenvalue problem

$$\left.\begin{aligned}
-\frac{\mathrm{d}}{\mathrm{d}x}\left(\alpha\frac{\mathrm{d}}{\mathrm{d}x}u\right) + \beta u = \lambda u(x) \quad (0 < x < \pi), \\
a_0 u(0) + b_0 u'(0) = a_1 u(\pi) + b_1 u'(\pi) = 0,
\end{aligned}\right\}$$

has a sequence of *simple* eigenvalues $\lambda_k = 1/\mu_k \to \infty$.

Another classical result we will need is the *Maximum Principle*.

Theorem 0.8   *Let $\Omega \subset \mathbb{R}^n$ be a bounded domain with smooth boundary and let $\lambda < \lambda_1$. If $u \in C^2(\Omega) \cup C(\overline{\Omega})$ is such that*

$$-\mathcal{L}u \geq \lambda u \text{ in } \Omega,$$

$$u \geq 0 \text{ on } \partial\Omega,$$

then $u \geq 0$ in Ω.

1

Differential calculus

This introductory chapter is mainly devoted to the differential calculus in Banach spaces. In addition to being a fundamental tool later on, the treatment of the calculus at this level permits better understanding at certain aspects, which might otherwise be neglected.

We discuss in Section 1 the Fréchet and Gâteaux derivatives as well as their elementary properties. The differentiability of the Nemitski operator is investigated in Section 2 and higher and partial derivatives are introduced in Sections 3 and 4, respectively.

1 Fréchet and Gâteaux derivatives

The Fréchet-differential is nothing else than the natural extension to Banach spaces of the usual definition of differential of a map in Euclidean spaces.

Let U be an open subset of X and consider a map $F : U \to Y$.

Definition 1.1 Let $u \in U$. We say that F is *(Fréchet-) differentiable* at u if there exists $A \in L(X, Y)$ such that, if we set

$$R(h) = F(u + h) - F(u) - A(h),$$

there results

$$R(h) = o(\|h\|), \tag{1.1}$$

that is

$$\frac{\|R(h)\|}{\|h\|} \to 0 \ \text{ as } \|h\| \to 0.$$

Such an A is uniquely determined and will be called the *(Fréchet) differential* of F at u and denoted by

$$A = dF(u).$$

If F is differentiable at all $u \in U$ we say that F is differentiable in U.

Hereafter, when there is no possible misunderstanding, *Fréchet differentiability* will be referred to simply as *differentiability*. A few comments on the preceding definition are in order.

(i)   Let us verify that A is unique. Supposing the contrary, let $B \in L(X, Y)$ satisfy Definition 1.1 and $A \neq B$. It follows that

$$\frac{\|Ah - Bh\|}{\|h\|} \to 0 \quad \text{as } \|h\| \to 0. \tag{1.2}$$

If $A \neq B$ there exists $h^* \in X$ such that $a := \|Ah^* - Bh^*\| \neq 0$. Taking $h = t h^*$, $t \in \mathbb{R} - \{0\}$, one has

$$\frac{\|A(th^*) - B(th^*)\|}{\|th^*\|} = \frac{\|Ah^* - Bh^*\|}{\|h^*\|} = \frac{a}{\|h^*\|},$$

a constant, in contradiction with (1.2).

(ii) If F is differentiable at u then

$$F(u + h) = F(u) + dF(u)h + o(\|h\|)$$

and F is continuous at the same point. Conversely if $F \in C(U, Y)$ then it is not necessary to require in Definition 1.1 the continuity of A. In fact (1.1) yields

$$A(h) = F(u + h) - F(u) - o(\|h\|)$$

and the continuity of F implies the continuity of A.

(iii) The definition of differentiability depends not on the norms but on the topology of X and Y only. That is if, for example, $\|.\|$ and $\||.\||$ are two equivalent norms on X then F is differentiable at u in $(X, \|.\|)$ if and only if F is in $(X, \||.\||)$ and the differential is the same.

Remark 1.2 The preceding comment (iii) could suggest the idea of extending the notion of Fréchet differentiability to locally convex topological spaces. The most natural way would be the following: let the topology of X (respectively Y) be induced by an infinite family of seminorms $|.|_{X,i}$ (resp. $\|.\|_{Y,j}$); define the differential of F as the linear continuous map A with the property that for all $|.|_{Y,j}$ there exists a seminorm $|.|_{X,i}$ such that $|F(u + h) - F(u) - Ah|_{Y,j} = o(|h|_{X,i})$. With such a definition all the main properties of the differential (below) hold true. Unfortunately, in dealing with the higher derivatives, there are strong difficulties and people introduced new classes of spaces, such as the "pseudotopological spaces", where a differential calculus suitable for the purposes of analysis can be carried out. These kind of topics, however, are beyond the purposes of our book.

Examples 1.3

(a)   The constant map $F(u) = c$ is differentiable at any u and $\mathrm{d}F(u) = 0$ for all $u \in X$.

(b)   Let $A \in L(X,Y)$. Since $A(u+h) - A(u) = A(h)$, it follows that A is differentiable in X and $\mathrm{d}A(u) = A$.

(c)   Let $B : X \times Y \to Z$ be a bilinear continuous map. There results
$$B(u + h, v + k) - B(u, v) = B(h, v) + B(u, k) + B(h, k).$$
From the continuity at the origin it follows that
$$\|B(h, k)\| \leq c\|h\|\,\|k\|.$$
Then B is differentiable at any $(u, v) \in X \times Y$ and $\mathrm{d}B(u, v)$ is the map $(h, k) \to B(h, v) + B(u, k)$.

(d)   Let H be a Hilbert space with scalar product $(.|.)$ and consider the map $F : u \to \|u\|^2 = (u|u)$. From
$$\|u + h\|^2 - \|u\|^2 = 2(u|h) + \|h\|^2$$
it follows that F is differentiable at any u and $\mathrm{d}F(u)h = 2(u|h)$. Note that $\|.\|$ is not differentiable at $u = 0$. For, otherwise, $\|h\| = Ah + o(\|h\|)$ for some $A \in L(H, \mathbb{R})$. Replacing h with $-h$ we would deduce that $\|h\| = -Ah + o(\|h\|)$ and hence $\|h\| = o(\|h\|)$, a contradiction.

(e)   If $X = \mathbb{R}$, $U = (a, b)$ and $F : U \to Y$ is differentiable at $t \in U$, the differential $\mathrm{d}F(t)$ can be identified with $\mathrm{d}F(t)[1] \in Y$ though the canonical isomorphism $i : L(\mathbb{R}, Y) \to Y$, $i(A) = A(1)$. For example, if $Y = \mathbb{R}^n$ and $F(t) = (f_i(t))_{i=1,\ldots,n}$, $\mathrm{d}F(t)$ "is" the vector with components $\mathrm{d}f_i/\mathrm{d}t$.

The main differentiation rules are collected in the following proposition.

Proposition 1.4

(i)   *Let $F, G : U \to Y$. If F and G are differentiable at $u \in U$ then $aF + bG$ is differentiable at u for any $a, b \in \mathbb{R}$ and*
$$\mathrm{d}(aF + bG)(u)h = a\,\mathrm{d}F(u)h + b\,\mathrm{d}G(u)h.$$

(ii)   *(Composite-map formula) Let $F : U \to Y$ and $G : V \to Z$ with*

$V \supset F(U)$, U *and* V *open subsets of* X *and* Y, *respectively, and consider the composite map*

$$G \circ F : U \to Z, \ G \circ F(u) := G(F(u)).$$

If F *is differentiable at* $u \in U$ *and* G *is differentiable at* $v := F(u) \in V$, *then* $G \circ F$ *is differentiable at* u *and*

$$d(G \circ F)(u)h = dG(v)[dF(u)h].$$

In other words the differential of $G \circ F$ *at* u *is the composition of the linear maps* $dF(u)$ *and* $dG(v)$, *with* $v = F(u)$.

The proofs of (i) and (ii) do not differ from those of the differentiation rules in $\mathbb{R}^n$.

Definition 1.5 Let $F : U \to Y$ be a differentiable in U. The map

$$F' : U \to L(X, Y), \ F' : u \to dF(u),$$

is called the *(Fréchet) derivative* of F.

If F' is continuous as a map from U to $L(X, Y)$ we will say that F is C^1 and write $F' \in C^1(U, Y)$.

Let us introduce the concept of *variational* (or *potential*) operator. If $Y = \mathbb{R}$, maps $J : U \to \mathbb{R}$ are usually called *functionals* and J' turns out to be a map from U to $L(X, \mathbb{R}) = X^*$ (the dual of X). In particular, if $X = H$ is a Hilbert space, $J'(u) \in H^*$ for all u and the Riesz Representation Theorem allows us to identify $J'(u)$ with an element of H. To be precise, we give the following definition.

Definition 1.6 Given a differentiable functional $J : U \to \mathbb{R}$ the *gradient* of J at u, denoted by $\nabla J(u)$, is the element of H defined by

$$(\nabla J(u)|h) = dJ(u)h, \ \text{for all } h \in H. \tag{1.3}$$

A map $F : U \to H$ with the property that there exists a differentiable functional $J : U \to \mathbb{R}$ such that $F = \nabla J$ is called a *variational* (or *potential*) *operator*.

As for maps in $\mathbb{R}^n$, we can also define here a directional derivative, usually called the Gâteaux differential (for short, G-differential).

Definition 1.7 Let $F : U \to Y$ be given and let $x \in U$. We say that F is *G-differentiable* at u if there exists $A \in L(X, Y)$ such that for all $h \in X$ there results

$$\frac{F(u + \varepsilon h) - F(u)}{\varepsilon} \to Ah \text{ as } \varepsilon \to 0. \tag{1.4}$$

The map A is uniquely determined, called the *G-differential* of F at u and denoted by $\mathrm{d_G} F(u)$.

Clearly, if F is Fréchet-differentiable at u then F is G-differentiable there and the two differentials coincide. Conversely, the G-differentiability does not imply the continuity of F, even: recall the elementary example $F : \mathbb{R}^2 \to \mathbb{R}^2$ defined by

$$F(s,t) = \left[\frac{s^2 t}{s^4 + t^2} \right]^2 \quad \text{if } t \neq 0,$$

$$F(s,0) = 0.$$

The following result replaces the elementary "Mean-Value Theorem" and plays a fundamental role in what follows.

For $u, v \in U$ let $[u,v]$ denote the segment $\{tu + (1-t)v : t \in [0,1]\}$.

Theorem 1.8   *Let $F : U \to Y$ be G-differentiable at any point of U. Given $u, v \in U$ such that $[u,v] \subset U$, there results*

$$\|F(u) - F(v)\| \leq \sup\{\|\mathrm{d_G} F(w)\| : w \in [u,v]\} \, \|u - v\|.$$

Proof. Without loss of generality we can assume that $F(u) \neq F(v)$. By a well-known corollary of the Hahn–Banach Theorem there exists $\psi \in Y^*$, $\|\psi\| = 1$, such that

$$\langle \psi, F(u) - F(v) \rangle = \|F(u) - F(v)\|. \tag{1.5}$$

Let $\gamma(t) = tu + (1-t)v$, $t \in [0,1]$, and consider the map $h : [0,1] \to \mathbb{R}$ defined by setting

$$h(t) = \langle \psi, F[\gamma(t)] \rangle = \langle \psi, F(tu + (1-t)v) \rangle.$$

From $\gamma(t + \tau) = \gamma(t) + \tau(u - v)$ it follows that

$$\frac{h(t+\tau) - h(t)}{\tau} = \left\langle \psi, \frac{F[\gamma(t+\tau)] - F[\gamma(t)]}{\tau} \right\rangle$$

$$= \left\langle \psi, \frac{F[\gamma(t) + \tau(u-v)] - F[\gamma(t)]}{\tau} \right\rangle. \tag{1.6}$$

Since F is G-differentiable in U, passing to the limit in (1.6) as $\tau \to 0$, we find

$$h'(t) = \langle \psi, \mathrm{d_G} F(tu + (1-t)v)(u - v) \rangle. \tag{1.7}$$

Applying the Mean–Value Theorem to h one has

$$h(1) - h(0) = h'(\theta) \quad \text{for some } \theta \in (0,1). \tag{1.8}$$

Substituting (1.5) and (1.7) into (1.8) we get

$$\|F(u) - F(v)\| = h(1) - h(0) = h'(\theta)$$
$$= \langle \psi, \mathrm{d}_G F(\theta u + (1 - \theta)v)(u - v) \rangle$$
$$\leq \|\psi\| \, \|\mathrm{d}_G F(\theta u + (1 - \theta)v)\| \, \|u - v\|.$$

Since $\|\psi\| = 1$ and $\theta u + (1 - \theta)v \in [u, v]$ the theorem follows.

As a consequence we can find a classic criterion of Fréchet differentiability.

Theorem 1.9 *Suppose* $F : U \to Y$ *is G-differentiable in* U *and let*

$$F'_G : U \to L(X, Y), \quad F'_G(u) = \mathrm{d}_G F(u),$$

be continuous at u^*. *Then* F *is Fréchet-differentiable at* u^* *and* $\mathrm{d}F(u^*)$
$= \mathrm{d}_G F(u^*)$.

Proof. We set

$$R(h) := F(u^* + h) - F(u^*) - \mathrm{d}_G F(u^*)h.$$

Plainly, R is G-differentiable in B_ε, for $\varepsilon > 0$ small enough, and

$$\mathrm{d}_G R(h) : k \to \mathrm{d}_G F(u^* + h)k - \mathrm{d}_G F(u^*)k. \qquad (1.9)$$

Applying Theorem 1.8 with $[u, v] = [0, h]$, we find (note that $R(0) = 0$)

$$\|R(h)\| \leq \sup_{0 \leq t \leq 1} \|\mathrm{d}_G R(th)\| \, \|h\|. \qquad (1.10)$$

From (1.9) with th instead of h, we deduce

$$\| \mathrm{d}_G R(th) \| = \|\mathrm{d}_G F(u^* + th) - \mathrm{d}_G F(u^*)\|.$$

Substituting into (1.10) we find

$$\|R(h)\| \leq \sup_{0 \leq t \leq 1} \|\mathrm{d}_G F(u^* + th) - \mathrm{d}_G F(u^*)\| \, \|h\|.$$

Since F'_G is continuous,

$$\sup_{0 \leq t \leq 1} \|\mathrm{d}_G F(u^* + th) - \mathrm{d}_G F(u^*)\| \to 0 \text{ as } \|h\| \to 0$$

and therefore $R(h) = o(\|h\|)$.

In view of Theorem 1.8, to find the Fréchet differential of F one can determine $\mathrm{d}_G F$ and show that F'_G is continuous.

Let $F \in C([a, b], X)$ and set (see subsection 0.3)

$$\Phi(t) = \int_a^t F(\xi)\mathrm{d}\xi.$$

It is immediately verifiable that Φ is differentiable and $\Phi'(t) = F(t)$

(we are using the canonical identification between $L(\mathbb{R}, X)$ and X; see Example 1.3 (e)). Φ is called a primitive of F.

From Theorem 1.8 it follows that

$$\| \Phi(t) - \Phi(s) \| \leq \sup \{\|F(\xi)\| \, (t-s) : s \leq \xi \leq t\}.$$

Hence, if $F(\xi) = 0$ for all $\xi \in [a, b]$, one has $\Phi =$ constant. In particular, Φ is, up to a constant, the unique primitive of F.

As a consequence we can obtain the following useful formula. Suppose that $[u, v] \in U$ and let $F \in C^1(U, Y)$. The map $F \circ \gamma : [0, 1] \to Y$, $F \circ \gamma(t) = F(tu + (1 - t)v)$ is C^1 and

$$(F \circ \gamma)'(t) = F'(tu + (1 - t)v) \, [u - v].$$

Integrating from 0 to 1 we get

$$F(v) - F(u) = \int_0^1 F'(tu + (1-t)v)[u - v]dt$$

$$= \left[\int_0^1 F'(tu + (1-t)v)dt \right] (u - v). \qquad (1.11)$$

Note that in the last integral F' is meant to take values in $L(X, Y)$.

2 Continuity and differentiability of Nemitski operators

In this section we want to study the differentiability of an important class of operators arising in nonlinear analysis: the so called "Nemitski operators" we are going to define.

Nemitski operators

Let Ω be an open bounded subset of $\mathbb{R}^n$ and let $M(\Omega)$ denote the class of real-valued functions $u : \Omega \to \mathbb{R}$ that are measurable on Ω. Here, and always hereafter, the measure is the Lebesgue one and will be denoted by μ; all the functions we will deal with in this section are taken in $M(\Omega)$.

Let $f : \Omega \times \mathbb{R} \to \mathbb{R}$ be given.

Definition 2.1 The *Nemitski operator* associated to f is the map defined on $M(\Omega)$ by setting

$$u(x) \to f(x, u(x)).$$

The same symbol f will be used to denote both f and its Nemitski operator.

We shall assume that f is a Carathéodory function. More precisely, we will say that f satisfies (C) if

(i) $s \to f(x,s)$ is continuous for almost every $x \in \Omega$,
(ii) $x \to f(x,s)$ is measurable for all $s \in \mathbb{R}$.

For the purpose of analysis it is particularly interesting when the Nemitski operators act on Lebesgue spaces $L^p = L^p(\Omega)$ (hereafter we will write L^p for $L^p(\Omega)$) and we shall discuss this case in some detail. Let us start by noticing that

$$f(u) \in M(\Omega) \text{ for all } u \in M(\Omega). \tag{2.1}$$

Indeed, if $u \in M(\Omega)$ there is a sequence χ_n of simple functions such that $\chi_n \to u$ a.e. in Ω. From (C) it follows that

$$f(\chi_n) \text{ is measurable and } f(\chi_n) \to f(u) \text{ a.e. in } \Omega,$$

and from this we deduce that $f(u) \in M(\Omega)$.

Continuity of Nemitski operators

Let $p, q \geq 1$ and suppose

$$|f(x,s)| \leq a + b|s|^\alpha, \quad \alpha = \frac{p}{q}, \tag{2.2}$$

for some constants $a, b > 0$.

Theorem 2.2 *Let $\Omega \subset \mathbb{R}^n$ be bounded and suppose f satisfies (C) and (2.2). Then the Nemitski operator f is a continuous map from L^p to L^q.*

For the proof we need the following measure-theoretic result (see, for example, [**Br**], Theorem IV.9).

Theorem 2.3 *Let $\mu(\Omega) < \infty$ and let $u_n \to u$ in L^p. Then there exist a sub-sequence u_{n_k} and $h \in L^p$ such that*

$$u_{n_k} \to u \quad a.e. \text{ in } \Omega, \tag{2.3}$$

$$|u_{n_k}| \leq h \quad a.e. \text{ in } \Omega. \tag{2.4}$$

Proof of Theorem 2.2 From (2.1)–(2.2) it follows immediately that $f(u) \in L^q$ whenever $u \in L^p$.
To show that f is continuous from L^p to L^q, let $u_n, u \in L^p$ be such that

$$\| u_n - u \|_{L^p} \to 0.$$

Using Theorem 2.3 we can find a sub-sequence $\{u_{n_k}\}$ of $\{u_n\}$ and $h \in L^p$

satisfying (2.3)–(2.4). Since u_{n_k} converges almost everywhere to u, it readily follows from (C) that

$$f(u_{n_k}) \to f(u) \text{ a.e. in } \Omega. \tag{2.5}$$

Moreover, from assumption (2.2) and (2.4) we infer

$$|f(u_{n_k})| \le a + b|u_{n_k}|^\alpha \le a + b|h|^\alpha \in L^q. \tag{2.6}$$

As an immediate consequence of the Lebesgue Dominated-Convergence Theorem, (2.5)–(2.6) yield

$$\|f(u_{n_k}) - f(u)\|_{L^q}^q = \int_\Omega |f(u_{n_k}) - f(u)|^q \to 0.$$

Since any sequence u_n converging to u in L^p has a sub-sequence u_{n_k} such that $f(u_{n_k}) \to f(u)$ in L^q, we can conclude that f is continuous at u, as a map from L^p to L^q.

Remark 2.4   Theorem 2.2 can be proved assuming that f satisfies (2.2) with a replaced by $a(x) \in L^q$.

Remark 2.5   It is possible to show that, if (C) holds and $f(u) \in L^q$ for all $u \in L^p$, then $f \in C(L^p, L^q)$. For this and other kinds of arguments we refer to [**Va**], p.154 and following.

Differentability of Nemitski operators

Our next result deals with the differentiability of Nemitski operators. First some remarks are in order.

Let $p > 2$ and suppose f has partial derivative $f_s = \partial f / \partial s$ satisfying (C) and such that

$$|f_s(x,s)| \le a + b|s|^{p-2} \tag{2.7}$$

for some constants $a, b > 0$. Since f_s satisfies (2.7), Theorem 2.2 applies and the Nemitski operator f_s is continuous from L^p to L^r, with $r = p/(p-2)$. As a consequence, for the function $f_s(u)v$ defined by

$$f_s(u)v : x \to f_s(x, u(x))v(x)$$

one has that $f_s(u)v \in L^{p'}$ for $u, v \in L^p$, where $p' = p/(p-1)$ is the conjugate exponent of p.

Theorem 2.6   *Let $\Omega \subset \mathbb{R}^n$ be bounded and suppose that $p > 2$ and f satisfies (C). Moreover, we suppose that $f(x, 0)$ is bounded and that f has partial derivative f_s satisfying (C) and (2.7).*
  Then $f : L^p \to L^{p'}$ is Fréchet-differentiable on L^p with differential

$$\mathrm{d}f(u) : v \to f_s(u)v.$$

Proof. Integrating (2.7) we find constants $c, d > 0$ such that
$$|f(x, s)| \leq c + d \, |s|^{p-1},$$
and another application of Theorem 2.2 yields the continuity of f as a map from L^p to $L^{p'}$, with $p' = p/(p-1)$.

For $u, v \in L^p$ we evaluate
$$\omega(u, v) = \|f(u+v) - f(u) - f_s(u)v\|_{L^{p'}}$$

$$= \left[\int_\Omega |f(x, u(x) + v(x)) - f(x, u(x)) - f_s(x, u(x))v(x)|^{p'} \right]^{1/p'}.$$

By the Mean-Value Theorem one has (for almost every $x \in \Omega$)
$$|f(x, u(x) + v(x)) - f(x, u(x)) - f_s(x, u(x))v(x)|$$

$$= |v(x) \int_0^1 [f_s(x, u(x) + \zeta v(x)) - f_s(x, u(x))] \mathrm{d}\zeta| = |v(x) \, w(x)|,$$

where

$$w(x) = \int_0^1 [f_s(x, u(x) + \zeta v(x)) - f_s(x, u(x))] \mathrm{d}\zeta.$$

With this notation and using the Hölder inequality we get that

$$\omega(u, v) = \left[\int_\Omega |v(x)w(x)|^{p'} \, \mathrm{d}x \right]^{1/p'}$$

$$\leq \|v\|_{L^p} \|w\|_{L^r} \quad \left(r = \frac{p}{p-2} \right). \tag{2.8}$$

Now, the norm $\|w\|_{L^r}$ can be estimated as follows:

$$\|w\|_{L^r}^r \leq \int_\Omega \mathrm{d}x \int_0^1 |f_s(x, u(x) + \zeta v(x)) - f_s(x, u(x))|^r \mathrm{d}\zeta$$

$$= \int_0^1 \mathrm{d}\zeta \int_\Omega |f_s(x, u(x) + \zeta v(x)) - f_s(x, u(x))|^r \mathrm{d}x$$

$$= \int_0^1 \|f_s(u + \zeta v) - f_s(u)\|_r^r \mathrm{d}\zeta. \tag{2.9}$$

As remarked before, f_s is continuous from L^p to L^r. Hence
$$\|f_s(u + \zeta v) - f_s(u)\|_{L^r}^r \to 0 \quad \text{as } \|v\|_{L^p} \to 0, \ \zeta \in [0, 1]. \tag{2.10}$$
From (2.8),(2.9) and (2.10) it follows that $\omega(u, v) = o(\|v\|_{L^p})$.

For $p = 2$ the above result does not hold, in general. Indeed, under the preceding assumptions, the Nemitski operator f is G-differentiable but, possibly, not Fréchet-differentiable. To be precise, let us assume that (C) holds for f, f_s and

$$|f_s(x, s)| \leq \text{const.} \qquad (2.11)$$

As before, it follows plainly that f is continuous from L^2 to L^2 and the map $v \to f_s(u)v$ from L^2 to L^2 is linear and bounded. In addition one has the following

Theorem 2.7 *Let $\Omega \subset \mathbb{R}^n$ be bounded and let f and f_s satisfy (C) and (2.11). Then $f : L^2 \to L^2$ is G-differentiable and $\mathrm{d}_G f(u)[v] = f_s(u)v$.*

Proof. According to Definition 1.7 we have to show that for all $u, v \in L^2$ there results

$$\left\| \frac{f(u + tv) - f(u)}{t} - f_s(u)v \right\|_{L^2} \to 0 \text{ as } t \to 0. \qquad (2.12)$$

As in the proof of theorem 2.6 one finds

$$\frac{f(u + tv) - f(u)}{t} - f_s(u)v = v \int_0^1 [f_s(u + \zeta tv) - f_s(u)]\mathrm{d}\zeta.$$

Letting

$$w_t = w_t(u, v) = \int_0^1 [f_s(u + \zeta tv) - f_s(u)]\mathrm{d}\zeta,$$

one has

$$\left\| \frac{f(u + tv) - f(u)}{t} - f_s(u)v \right\|_{L^2}^2 = \int_\Omega v^2 w_t^2$$

$$\leq \int_\Omega v^2 \mathrm{d}x \int_0^1 |f_s(u + \zeta tv) - f_s(u)|^2 \mathrm{d}\zeta.$$

When $t \to 0$ then $t\zeta v \to 0$ a.e. in Ω and hence

$$f_s(u + t\zeta v) - f_s(u) \to 0 \text{ a.e. in } \Omega.$$

Since

$$|f_s(x, u(x) + t\zeta v(x)) - f_s(x, u(x))|^2 \leq \text{const.}$$

the Lebesgue Dominated-Convergence Theorem implies

$$\int_0^1 |f_s(u + \zeta tv) - f_s(u)|^2 \mathrm{d}\zeta \to 0, \text{ as } t \to 0, \qquad (2.13)$$

and (2.12) follows.

The preceding theorem is completed by the following proposition.

Proposition 2.8 *Let $\Omega \subset \mathbb{R}^n$ be bounded, suppose f and f_s satisfy* (C) *and* (2.11) *and let f be Fréchet-differentiable at some $u^* \in L^2$. Then there exists $a(x), b(x) \in M(\Omega)$ such that*

$$f(x, u) = a(x)u + b(x).$$

Proof. Suppose first that $u^* = 0$ and $f(x, 0) = 0$.

Let $D(y, \delta)$ denote the ball centred at $y \in \Omega$ with measure δ. Given $x^* \in \Omega$ and $\lambda \in \mathbb{R}$, consider the functions $v_\delta(x) \in L^2(\Omega)$ given by

$$v_\delta(x) = \lambda, \text{ for } x \in D(x^*, \delta),$$

$$v_\delta(x) = 0, \text{ for } x \in \Omega \backslash D(x^*, \delta).$$

Obviously $v_\delta \to 0$ in L^2 as $\delta \to 0$.

Recall that if f is Fréchet-differentiable at $u^* = 0$ then $f'(0) = \mathrm{d}_G f(0)$. Hence $f'(0)v = f_s(0)v$, and therefore

$$\frac{\|f(v_\delta) - f_s(0)v_\delta\|_{L^2}}{\|v_\delta\|_{L^2}} \to 0, \text{ as } \delta \to 0.$$

By a direct calculation one finds

$$\frac{\|f(v_\delta) - f_s(0)v_\delta\|_{L^2}}{\|v_\delta\|_{L^2}} = \frac{1}{|\lambda|\sqrt{\delta}} \left[\int_{D(x^*, \delta)} |f(x, \lambda) - f_s(x, 0)\lambda|^2 \right]^{1/2}$$

and hence

$$\frac{1}{|\lambda|\sqrt{\delta}} \left[\int_{D(x^*, \delta)} |f(x, \lambda) - f_s(x, 0)\lambda|^2 \right]^{1/2} \to 0, \text{ as } \delta \to 0. \qquad (2.14)$$

If we set

$$\psi_\lambda(x) = \left| \frac{f(x, \lambda) - f_s(x, 0)\lambda}{\lambda} \right|^2,$$

(2.14) becomes

$$\frac{1}{\delta} \int_{D(x^*, \delta)} \psi_\lambda(x)\mathrm{d}x \to 0, \text{ as } \delta \to 0. \qquad (2.15)$$

Since for all $\psi \in L^1$ and almost every $y \in \Omega$ there results

$$\frac{1}{\delta} \int_{D(y, \delta)} \psi(x)\mathrm{d}x \to \psi(y), \text{ as } \delta \to 0,$$

we deduce from (2.15) that for all $\lambda \in \mathbb{R}$ there exists a null set N_λ such that for all $y \notin N_\lambda$ one has $\psi_\lambda(y) = 0$.

Taking λ in a countable dense subset Λ of $\mathbb{R}$ and letting $N = \cup_{\lambda \in \Lambda} N_\lambda$ one infers that for all $y \notin N$ and all $\lambda \in \Lambda$ there results $\psi_\lambda(y) = 0$ that is,

$$f(y, \lambda) = f_s(y, 0)\lambda. \tag{2.16}$$

Using the continuity of $f(x, .)$, one deduces that (2.16) holds true for all $\lambda \in \mathbb{R}$ and almost every y.

Lastly, let us set

$$g(x, u) = f(x, u + u^*) - f(u^*).$$

One has that g is Fréchet-differentiable at 0 and $g(x, 0) = 0$. Applying the preceding arguments to g, we find

$$f(x, u + u^*) - f(u^*) = f_s(x, u^*)u$$

and the proposition follows.

Potential operators

We end this section by dealing with potential operators (Definition 1.6). The results will not be used in the remainder of this book but are important in connection with variational problems.

Let $H_0^1 = H_0^1(\Omega)$ (where Ω is a bounded domain of $\mathbb{R}^n$) denote the usual Sobolev space (see Subsection 0.5) with scalar product $(.|.)_{H^{1,2}}$ and norm $\|.\|_{H^{1,2}}$. Let $n > 2$ (if $n = 1, 2$, the arguments we are going to expand apply as well, with some modifications; see Remark 2.10 below). Suppose f satisfies (C) and

$$|f(x, s)| \le a + b|s|^\sigma \quad \text{with } \sigma \le \frac{n+2}{n-2} = 2^* - 1, \tag{2.17}$$

Here $2^* = 2n/(n-2)$ and, by the Sobolev Embedding Theorem (see Theorem 0.3 (i)), $H_0^1 \hookrightarrow L^{2^*}$ and

$$\|v\|_{L^{2^*}} \le \text{const.} \|v\|_{H^{1,2}}$$

By Theorem 2.2 it follows that

$$f \in C(L^{2^*}, L^q) \quad \text{with } q = \frac{2^*}{\sigma} \ge \frac{2n}{n+2}. \tag{2.18}$$

In particular one has $f(u) \in L^{2n/(n+2)}$ for all $u \in H_0^1$. As a consequence, $f(u)v \in L^1$ for all $u, v \in H_0^1$ and the equality

$$(N(u)|v)_{H^{1,2}} = \int_\Omega f(x, u(x))v(x)\mathrm{d}x, \quad u, v \in H_0^1, \tag{2.19}$$

defines an operator $N : H_0^1 \to H_0^1$. Note that N is continuous. To see this we evaluate

$$\|N(u) - N(v)\|_{H^{1,2}} = \sup \left\{ \left| \int_\Omega [f(x, u) - f(x, v)]w\mathrm{d}x \right| : \|w\|_{H^{1,2}} \le 1 \right\}$$

$$\leq \sup\{\|f(u) - f(v)\|_{L^{2n/(n+2)}} \|w\|_{L^{2^*}} : \|w\|_{H^{1,2}} \leq 1\}$$

$$\leq c\|f(u) - f(v)\|_{L^{2n/(n+2)}}.$$

If $u_m \to v$ in H_0^1 one has (by the Sobolev Embedding Theorem) $u_m \to v$ in L^{2^*} and (by (2.18)) $f(u_m) \to f(v)$ in $L^{2n/(n+2)}$.

Set

$$F(x, s) = \int_0^s f(x, t)\mathrm{d}t.$$

From (2.17) it follows that there exist $c, d > 0$ such that

$$|F(x, s)| \leq c + d|s|^{2^*}. \tag{2.20}$$

Then $F(., u(.)) \in L^1$ for all $u \in H_0^1 (\subset L^{2^*})$ and it makes sense to define a functional $\phi : H_0^1 \to \mathbb{R}$ by setting

$$\phi(u) = \int_\Omega F(x, u(x))\mathrm{d}x.$$

The functional ϕ can be obtained by composition according to the following diagram

$$H_0^1 \xrightarrow{\alpha} L^{2^*} \xrightarrow{F} L^{2n/(n+2)} \xrightarrow{\beta} L^1 \rightarrow \mathbb{R}$$

$$u \rightarrow u \rightarrow F(., u(.)) \rightarrow F(., u(.)) \rightarrow \int_\Omega F(x, u(x))\mathrm{d}x$$

where α and β stand for the embedding of H_0^1 into L^{2^*}, of $L^{2n/(n+2)}$ into L^1 respectively. Since F satisfies (2.20), Theorem 2.6 applies to F as a map from L^{2^*} to $L^{2n/(n+2)}$. From Proposition 1.4(b) (derivative of the composite map) it follows that ϕ is differentiable with

$$\phi'(u) : v \longrightarrow \int_\Omega f(x, u(x))v(x)\mathrm{d}x.$$

Then, recalling the definition of "gradient" (Definition 1.6), we have the following

Theorem 2.9   *Let $\Omega \subset \mathbb{R}^n$ be bounded and suppose f satisfies (C) and (2.17). Then ϕ is a C^1 functional on H_0^1 with gradient*

$$\nabla\phi(u) = N(u),$$

where N is defined in (2.19).

Remark 2.10   If $n = 2$ then the same result holds assuming f satisfies (2.17) with any $\sigma < \infty$. It suffices to repeat the above arguments using the stronger form of the Sobolev Embeddings when $\Omega \subset \mathbb{R}^2$.

3 Higher derivatives

Let $F \in C(U,Y)$ be differentiable in the open set $U \subset X$ and consider $F' : U \to L(X,Y)$

Definition 3.1 Let $u^* \in U : F$ is *twice (Fréchet-) differentiable* at u^* if F' is differentiable at u^*. The *second (Fréchet) differential* of F at u^* is defined as

$$\mathrm{d}^2 F(u^*) = \mathrm{d}F'(u^*).$$

If F is twice differentiable at all points of U we say that F is twice differentiable in U.

According to the above definition $\mathrm{d}^2 F(u^*)$ is a linear continuous map from X to $L(X,Y)$:

$$\mathrm{d}^2 F(u^*) \in L(X, L(X,Y)).$$

It is convenient to see $\mathrm{d}^2 F(u^*)$ as a bilinear map on X. For this, let $L_2(X,Y)$ denote the space of continuous bilinear maps from $X \times X \to Y$. To any $A \in L(X, L(X,Y))$ we can associate $\Phi_A \in L_2(X,Y)$ given by $\Phi_A(u_1, u_2) = [A(u_1)](u_2)$. Conversely, given $\Phi \in L_2(X,Y)$ and $h \in X$, $\Phi(h,.) : k \to \Phi(h,k)$ is a continuous linear map from X to Y; hence to any $\Phi \in L_2(X,Y)$ is associated the linear application $X \to L(X,Y)$,

$$\tilde{\Phi} : h \to \Phi(h,.) \in L(X,Y).$$

It is easy to see that in this way we define an isomorphism between $L(X, L(X,Y))$ and $L_2(X,Y)$. Actually, such an isomorphism is an isometry because there results

$$\|\tilde{\Phi}\|_{L(X,L(X,Y))} = \sup_{\|h\| \leq 1} \|\tilde{\Phi}(h)\|_{L(X,Y)}$$

$$= \sup_{\|h\| \leq 1} \sup_{\|k\| \leq 1} \|\Phi(h,k)\| = \|\Phi\|_{L_2(X,Y)}.$$

In the following we will use the same symbol $\mathrm{d}^2 F(u^*)$ to denote the continuous bilinear map obtained by the preceding isometry. The value of $\mathrm{d}^2 F(u^*)$ at a pair (h,k) will be denoted by

$$\mathrm{d}^2 F(u^*)[h,k].$$

If F is twice differentiable in U, the *second (Fréchet) derivative* of F is the map $F'' : U \to L_2(X,Y)$,

$$F'' : u \to \mathrm{d}^2 F(u).$$

If F'' is continuous from U to $L_2(X,Y)$ we say that $F \in C^2(U,Y)$.

Examples 3.2

(i)   If $A \in L(X, Y)$ then $A \in C^2(X, Y)$ and $\mathrm{d}^2 A[h, k] = 0$ for all
       $(h, k) \in X \times X$.
(ii)   Let $X = C([0, 1])$ and $F : X \to X$, $F : u(t) \to u^2(t)$. $F \in$
       $C^2(X, X)$ and
$$\mathrm{d}^2 F(u) : (h(t), k(t)) \to 2h(t)k(t).$$

The following proposition can be useful for evaluating $\mathrm{d}^2 F(u)$.

Proposition 3.3   *Let $F : U \to Y$ be twice differentiable at $u^* \in U$.
Then for all fixed $h \in X$ the map $F_h : X \to Y$ defined by setting*
$$F_h(u) = \mathrm{d}F(u)h$$
is differentiable at u^* and $\mathrm{d}F_h(u^*)k = F''(u^*)[h, k]$.

Proof. F_h is obtained by composition
$$U \xrightarrow{\mathrm{d}F} L(X, Y) \xrightarrow{\mathcal{E}_h} Y$$
$$u \longrightarrow \mathrm{d}F(u) \longrightarrow \mathrm{d}F(u)h$$
between the derivative $u \to \mathrm{d}F(u)$ and the "evaluation map" $\mathcal{E}_h$ which
associates to each $A \in L(X, Y)$ the value $A(h) \in Y$. Since $\mathcal{E}_h$ is linear,
the result follows by the composite mapping formula 1.4-(ii).

We have seen that $F''(u)$ can be regarded as a bilinear map. More
precisely one has the following

Theorem 3.4   *If $F : U \to Y$ is twice differentiable at $u \in U$, then
$F''(u) \in L_2(X, Y)$ is symmetric.*

Proof. For $h, k \in X$ with $h, k \in B(\varepsilon)$ (ε small enough), we set
$$\psi(h, k) = F(u + h + k) - F(u + h) - F(u + k) + F(u),$$
$$\gamma_h(\xi) = F(u + h + \xi) - F(u + \xi),$$
and consider, for h fixed, the map $g_h : B(\varepsilon) \to Y$,
$$g_h : k \to \psi(h, k) - F''(u)[h, k] = \gamma_h(k) - \gamma_h(0) - F''(u)[h, k].$$
Since F is differentiable in U and $F''(u)(h) : k \to F''(u)[h, k]$ is linear
(as a map from X to $L(X, Y)$), Theorem 1.8 yields
$$\|\psi(h, k) - F''(u)[h, k]\|$$
$$\leq \sup\{\|\mathrm{d}\gamma_h(tk) - F''(u)(h)\| : 0 \leq t \leq 1\}\| k\|$$
$$= \sup\{\|\mathrm{d}F(u + h + tk) - \mathrm{d}F(u + tk)$$
$$- F''(u)(h)\| : 0 \leq t \leq 1\}\|k\|. \tag{3.1}$$

Since F is twice differentiable at $u \in U$, one has

$$F'(u + h + tk) = F'(u) + F''(u)(h + tk) + \omega(h + tk),$$
$$F'(u + tk) = F'(u) + F''(u)(tk) + \omega(tk),$$

with $\omega(v) = o(\|v\|)$. Hence

$$F'(u + h + tk) - F'(u + tk) = F''(u)(h) + \omega(h + tk) - \omega(tk). \quad (3.2)$$

Using (3.1) and (3.2) and taking into account that $\omega(v) = o(\|v\|)$ we get, that

$$\|\psi(h, k) - F''(u)[h, k]\| \leq \sup\{\|\omega(h + tk) - \omega(tk)\| : 0 \leq t \leq 1\}\|k\|$$
$$\leq \varepsilon(\|h\| + 2\|k\|)\,\|k\|, \quad (3.3)$$

provided $\|h\|$ and $\|k\|$ are sufficiently small.

Exchanging the roles of h, k we get (for $\|h\|, \|k\|$ small)

$$\|\psi(k, h) - F''(u)[k, h]\| \leq \sup\{\|\omega(k + th) - \omega(th)\| : 0 \leq t \leq 1\}\,\|h\|$$
$$\leq \varepsilon(\|k\| + 2\|h\|)\,\|h\|. \quad (3.4)$$

Since $\psi(h, k) = \psi(k, h)$ we deduce from (3.3) and (3.4)

$$\|F''(u)[h, k] - F''(u)[k, h]\| \leq \varepsilon(2\|k\|^2 + 2\|h\|^2 + 2\|h\|\,\|k\|)$$
$$\leq 3\,\varepsilon(\|k\|^2 + \|h\|^2). \quad (3.5)$$

Inequality (3.5) has been proved for $\|h\|, \|k\|$ small enough, but holds true for all $\|h\|, \|k\|$, because $F''(u)[h, k]$ is homogeneous of degree 2. Since ε is arbitrary, (3.5) implies that $F''(u)[h, k] = F''(u)[k, h]$ for all h, k.

To define $(n + 1)$-th derivatives $(n \geq 2)$ we can proceed by induction.

Given $F : U \to Y$, let F be n times differentiable in U. The nth differential at a point $x \in U$ will be identified with a continuous n-linear map from $X \times X \times \ldots \times X$ (n times) to Y (recall that, as before, there is an isometry between $L(X, \ldots, L(X, Y))\ldots$ and $L_n(X, Y)$).

Let $F^{(n)} : U \to L_n(X, Y)$ denote the map

$$F^{(n)} : u \to \mathrm{d}^n F(u).$$

The $(n + 1)$-th differential at u^* will be defined as the differential of $F^{(n)}$, namely

$$\mathrm{d}^{(n+1)} F(u^*) = \mathrm{d}F^{(n)}(u^*) \in L(X, L_n(X, Y)) \approx L_{n+1}(X, Y).$$

We will say that $F \in C^n(U, Y)$ if F is n times (Fréchet) differentiable in U and the nth derivative $F^{(n)}$ is continuous from U to $L_n(X, Y)$. The value of $\mathrm{d}^n F(u^*)$ at $(h_1, \ldots, h_n)$ will be denoted by

$$\mathrm{d}^n F(u^*)[h_1, \ldots, h_n].$$

If $h = h_1 = \ldots = h_n$ we will write for short $\mathrm{d}^n F(u^*)[h]^n$.

In order to extend Theorem 3.4 to higher derivatives some preliminaries are in order. Given a map $G : U \to L_m(X, Y)$ and the point
$\mathbf{h} = (h_1, \ldots, h_m) \in X \times \ldots \times X$, we can associate G with the map
$G[\mathbf{h}] : U \to Y$ defined by setting

$$G[\mathbf{h}](u) = G(u)[h_1, \ldots, h_n].$$

We can immediately see (see Proposition 3.3) that if G is differentiable
at u then $G[\mathbf{h}]$ is differentiable at u and there results

$$d(G[\mathbf{h}])(u) : v \to dG(u)[v, h_1, \ldots, h_n]. \tag{3.6}$$

Let F be n times differentiable on U and set $\mathbf{h} = (h_2, \ldots, h_n)$. Applying (3.6) to $G = d^{n-1}F$, we find that

$$d(d^{n-1}F[\mathbf{h}])(u)[h_1] = d^n F(u^*)[h_1, \ldots, h_n]. \tag{3.7}$$

Theorem 3.5 *If $F : U \to Y$ is n times differentiable in U, then the
map*

$$(h_1, \ldots, h_n) \to d^n F(u^*)[h_1, \ldots, h_n]$$

is symmetric.

Proof. The result is true for $n = 2$ (Theorem 3.4). By induction on n,
let the claim hold for $n - 1 \geq 2$. Then

$$d^{n-1}F(u)[h_2, \ldots, h_i, \ldots, h_j, \ldots, h_n]$$
$$= d^{n-1}F(u)[h_2, \ldots, h_j, \ldots, h_i, \ldots, h_n].$$

Applying (3.7) to $h(u) = d^{n-1}F(u)[h_2, \ldots, h_i, \ldots, h_j, \ldots, h_n]$ we get
that

$$d^n F(u^*)[h_1, h_2, \ldots, h_i, \ldots, h_j, \ldots, h_n]$$
$$= d^n F(u^*)[h_1, h_2, \ldots, h_j, \ldots, h_i, \ldots, h_n]. \tag{3.8}$$

Similarly, letting $G(u) = d^{n-2}F(u)[h_3, \ldots, h_n]$, one has

$$d^2 G(u^*)[h_1, h_2] = d^n F(u^*)[h_1, h_2, h_3, \ldots, h_n],$$

and from Theorem 3.4 it follows that

$$d^n F(u^*)[h_1, h_2, h_3, \ldots, h_n] = d^2 G(u^*)[h_1, h_2]$$
$$= d^2 G(u^*)[h_2, h_1] = d^n F(u^*)[h_2, h_1, h_3, \ldots, h_n]. \tag{3.9}$$

The symmetry of $d^n F(u^*)$ is an immediate consequence of (3.8) and
(3.9).

4 Partial derivatives, Taylor's formula

Let us consider two Banach spaces X, Y and let $(u^*, v^*) \in X \times Y$. Define

mappings $\sigma_{v^*} : X \to X \times Y$ and $\tau_{u^*} : Y \to X \times Y$ as follows.
$$\sigma_{v^*}(u) = (u, v^*);$$
$$\tau_{u^*}(y) = (u^*, v).$$

Notice that the derivatives of σ_{v^*} and τ_{u^*} are respectively, the linear maps
$$\sigma := \mathrm{d}\sigma_{v^*} : h \to (h, 0),$$
$$\tau := \mathrm{d}\tau_{u^*} : k \to (0, k).$$

Let Q be an open subset of $X \times Y$, $(u^*, v^*) \in Q$ and $F : Q \to Z$.

Definition 4.1 If the map $F \circ \sigma_{v^*}$ is differentiable at u^* we say that F is *differentiable with respect to u* at (u^*, v^*). The linear map $\mathrm{d}[F \circ \sigma_{v^*}](u^*) \in L(X, Z)$ is called the *partial derivative* of F at (u^*, v^*) with respect to u and denoted by $\mathrm{d}_u F(u^*, v^*)$.

Similarly, if $F \circ \tau_{u^*}$ is differentiable at v^* we say that F is *differentiable with respect to v* at (u^*, v^*) and the linear map $\mathrm{d}[F \circ \tau_{u^*}](v^*) \in L(Y, Z)$ is called the *v-partial derivative* of F at (u^*, v^*) and denoted by $\mathrm{d}_v F(u^*, v^*)$.

The preceding definition is equivalent to requiring that there exist a linear map $A_u \in L(X, Z)$ (resp. $A_v \in L(Y, Z)$), such that
$$F(u^* + h, v^*) - F(u^*, v^*) = A_u(h) + o(\|h\|),$$
$$F(u^*, v^* + k) - F(u^*, v^*) = A_v(k) + o(\|k\|).$$

The following result is an immediate consequence of Definition 4.1 and the differentation rule 1.4 (ii).

Proposition 4.2 *If F is differentiable at (u^*, v^*) then F has partial derivatives with respect to u and v at (u^*, v^*) and we have*
$$\mathrm{d}_u F(u^*, v^*)(h) = \mathrm{d}F(u^*, v^*)\sigma(h) = \mathrm{d}F(u^*, v^*)(h, 0),$$
$$\mathrm{d}_v F(u^*, v^*)(k) = \mathrm{d}F(u^*, v^*)\tau(k) = \mathrm{d}F(u^*, v^*)(0, k).$$

In quite similar way one can define higher partial derivatives. For example, if F has u-partial derivative at all $(u, v) \in Q$, we can define the map $F_u : Q \to L(X, Z)$ by setting
$$F_u(u, v) = \mathrm{d}_u F(u, v).$$

Then the partial derivative $\mathrm{d}_{u,v} F(u^*, v^*)$ is the v-derivative at (u^*, v^*) of F_u, namely
$$\mathrm{d}_{u,v} F(u^*, v^*) = \mathrm{d}_v[F_u](u^*, v^*).$$

The map $F_{u,v} : Q \to L(Y, L(X, Z))$ will be defined by setting
$$F_{u,v}(u, v) = \mathrm{d}_{u,v} F(u, v).$$

Moreover, if F is twice differentiable at (u^*, v^*), then $d_{u,v}F(u^*, v^*)$ is the bilinear map from $X \times Y$ to Z given by

$$(h, k) \to F''(u^*, v^*)[\sigma h, \tau k]. \tag{4.1}$$

The notation $d^m_{u^\ell, v^{m-\ell}}$ will be employed to indicate

$$d^m_{u^\ell, v^{m-\ell}} = d^\ell_{u^\ell} \left(d^{m-\ell}_{v^{m-\ell}} \right).$$

The definition of partial derivative given above permits us to obtain in a rather straightforward way all the classical results of calculus.

For example one can prove the following.

Theorem 4.3 *Suppose that*

(i)   *F has u- and v-derivatives in a neighbourhood N of $(u^*, v^*) \in Q$,*
(ii)   *F_u and F_v are continuous in N.*
Then F is differentiable at (u^*, v^*).

As another example, we can use (4.1) and Theorem 3.4 to show

$$d_{u,v}F(u^*, v^*)[h, k] = F''(u^*, v^*)[\sigma h, \tau k]$$
$$= F''(u^*, v^*)[\tau k, \sigma h] = d_{v,u}F(u^*, v^*)[k, h],$$

which is nothing else than the classical Schwarz Theorem.

Taylor's formula

Let $F \in C^n(Q, Y)$ and let $u, u + v \in Q$ be such that the interval $[u, u + v] \subset Q$.

Set $\gamma(t) = u + tv$, $t \in [0, 1]$ and let $\phi : [0, 1] \to Y$ be defined by

$$\phi(t) = F(\gamma(t)).$$

Using Proposition 1.4 (ii) and (3.7) it follows readily that the function ϕ is C^n and there result

$$\phi'(t) = dF(u + tv)[v],$$
$$\phi''(t) = d^2F(u + tv)[v]^2,$$

$$\dotsi\dotsi\dotsi\dotsi\dotsi$$

$$\phi^{(n)}(t) = d^n F(u + tv)[v]^n.$$

By elementary calculations one has

$$\phi(1) = \phi(0) + \phi'(0) + \frac{1}{2!}\phi''(0) + \cdots + \frac{1}{(n-1)!}\phi^{(n)}(0)$$

$$+ \frac{1}{(n-1)!} \int_0^1 (1-t)^{n-1}\phi^{(n)}(t)dt,$$

and hence
$$F(u + v) = F(u) + dF(u)[v] + \cdots$$

$$+ \frac{1}{(n-1)!} \int_0^1 (1-t)^{n-1} d^{(n)} F(u + tv)[v]^n dt.$$

The last integral can be written as

$$\frac{1}{(n-1)!} \int_0^1 (1-t)^{n-1} d^{(n)} F(u + tv) dt [v]^n$$

$$= \frac{1}{n!} d^n F(u)[v]^n + \varepsilon(u, v)[v]^n, \tag{4.2}$$

where

$$\varepsilon(u, v) = \frac{1}{(n-1)!} \int_0^1 (1-t)^{n-1} [d^{(n)} F(u+tv) - d^{(n)} F(u)] dt \to 0 \text{ as } v \to 0.$$

Lastly, let us write explicitly the form of (4.2) when $F = F(u, v)$ is defined on $Q \subset X \times Y$ with values in Z and is C^n, that is, has continuous partial derivatives up to order n. We write (u, v) instead of u and set $w = (h, k) = \sigma h + \tau k$. If we use Proposition 4.2 the mth term in (4.2) becomes

$$\frac{1}{m!} d^{(m)} F(u, v)[w]^m = \frac{1}{m!} d^{(m)} F(u, v)[\sigma h + \tau k]^m$$

$$= \frac{1}{m!} d^{(m)} F(u, v) \sum \binom{m}{\ell} [\sigma h]^\ell [\tau k]^{m-\ell}$$

$$= \frac{1}{m!} \sum \binom{m}{\ell} d^{(m)} F(u, v) [\sigma h]^\ell [\tau k]^{m-\ell}$$

$$= \frac{1}{m!} \sum \binom{m}{\ell} d^m_{u^\ell, v^{m-\ell}} F(u, v) [h]^\ell [k]^{m-\ell}.$$

Remark 4.4 (on notation) Hereafter we will often deal with maps $F : \mathbb{R} \times X \to Y$ depending on a real parameter λ. In such a case the mixed derivative $F_{u,\lambda}(\lambda_o, u_o)$ is a linear map from $\mathbb{R}$ to $L(X, Y)$: $F_{u,\lambda}(\lambda_o, u_o) \in L(\mathbb{R}, L(X, Y))$. Then, in accordance with what we remarked in Example 1.3 (e), we can and will identify $F_{u,\lambda}(\lambda_o, u_o)$ with the linear map $h \to F_{u,\lambda}(\lambda_o, u_o)[h, 1]$.

2

Local inversion theorems

This chapter deals with the local inversion of maps between Banach spaces. The first section contains a general inversion result; in the second maps depending on a parameter are investigated. An application to the stability of orbits is given in Section 3.

1 The Local Inversion Theorem

For simplicity of notation we will deal with maps $F \in C(X, Y)$ where X, Y are Banach spaces; maps defined on an open subset of X could be treated with minor changes only.

Let us start with some preliminaries. Let $A \in L(X, Y)$. A is invertible if there exists $B \in L(Y, X)$ such that

$$B \circ A = I_X,$$
$$A \circ B = I_Y$$

The map B is obviously unique and will be denoted by A^{-1}. We also set

$$\text{Inv}(X, Y) = \{A \in L(X, Y) : A \text{ is invertible }\}.$$

Let us recall for future reference that, as a consequence of the "Closed Graph Theorem", if $A \in L(X, Y)$ is injective and $R(A) = Y$, then $A \in \text{Inv}(X, Y)$.

The following result is also well known (see, for example, [**Fi**], 3.1).

Proposition 1.1

(i) $\mathrm{Inv}(X,Y)$ *is an open subset of* $L(X,Y)$. *More precisely, if* $A \in$ $\mathrm{Inv}(X,Y)$ *then any* $T \in L(X,Y)$ *such that*

$$\|T - A\| < \frac{1}{\|A^{-1}\|}$$

is invertible.

(ii)   *The map* $J : \mathrm{Inv}(X,Y) \to L(Y,X)$ *defined by* $J(A) = A^{-1}$, *is* C^k *for all* $k \geq 1$ *(i.e.* C^∞).

It can be remarked that the continuity of $J' : A \to \mathrm{d}J(A)$ can be deduced directly by the fact that J is differentiable and $\mathrm{d}J(A)[B] = -A^{-1} \circ B \circ A^{-1}$.

Let U (resp. V) be an open subset of X (resp. Y). We say that $F \in \mathrm{Hom}(U,V)$ if there exists a map $G : V \to U$, such that

$$G(F(u)) = u \quad \text{for all } u \in U, \tag{1.1}$$

$$F(G(v)) = v \quad \text{for all } v \in V. \tag{1.2}$$

The map $F \in C(X,Y)$ is said to be *locally invertible* at $u^* \in X$ if there exist neighbourhoods U of u^* and V of $v^* = F(u^*) \in Y$ such that $F \in \mathrm{Hom}(U,V)$.

In other words F is locally invertible at u^* if there are neighbourhoods U and V of u^* and y^*, respectively, and a map $G : V \to U$ satisfying (1.1)–(1.2). The map G is called the (local) inverse of F and denoted by F^{-1}.

The following properties are direct consequences of the definition.

(a)   (*Transitivity*) If $F_1 \in C(X_1, X_2)$ is locally invertible at u and $F_2 \in C(X_2, Y)$ is locally invertible at $v = F_1(u)$, then $F_2 \circ F_1$ is locally invertible at u.

(b)   (*Stability*) If F is locally invertible at u, then it is locally invertible at any point in a suitable neighbourhood of u.

Let F be locally invertible at u^* and suppose F and $G = F^{-1}$ are differentiable at u^* and v^*, respectively. Differentiating (1.1) (resp. (1.2)) at u^* (resp. v^*) we find

$$G'(v^*) \circ F'(u^*) = I_X,$$

$$F'(u^*) \circ G'(v^*) = I_Y,$$

and this means that $F'(u^*) \in \mathrm{Inv}(X,Y)$ with inverse $G'(v^*) \in \mathrm{Inv}(Y,X)$.

The following theorem states that, under suitable assumptions, the converse is also true.

Theorem 1.2 (Local Inversion Theorem)   *Suppose $F \in C^1(X, Y)$ and $F'(u^*) \in \text{Inv}(X, Y)$. Then F is locally invertible at u^* with C^1 inverse. More precisely, there exist neighbourhoods U of u^* and V of $v^* = F(u^*)$ such that*

(i)   *$F \in \text{Hom}(U, V)$,*
(ii)   *$F^{-1} \in C^1(V, X)$ and for all $v \in V$ there results*
$$\mathrm{d}F^{-1}(v) = (F'(u))^{-1}, u = F^{-1}(v), \tag{1.3}$$
(iii)   *if $F \in C^k(X, Y), k > 1$, then $F^{-1} \in C^k(V, X)$.*

Proof. (i) Up to a translation, there is no loss of generality if we take $u^* = 0$ and $v^* = F(0) = 0$. Moreover, according to the transitivity property, it suffices to discuss the local invertibility of $A \circ F$, where A is any linear invertible map. With the choice

$$A = [F'(0)]^{-1}$$

we are led to consider the case in which

$$F = I + \Psi, \text{ where } \Psi \in C^1(X, X) \text{ and } \Psi'(0) = 0.$$

Here $I = I_X$ denotes the identity on X. Let $r > 0$ be such that

$$\|\Psi'(p)\| < \frac{1}{2}, \text{ for all } \|p\| < r. \tag{1.4}$$

Using Theorem 1.1.8 we find for all $p, q \in B(r)$

$$\|\Psi(p) - \Psi(q)\| \le \sup\{\|\Psi'(w)\| : w \in [p, q]\}\|p - q\|$$

$$\le \frac{1}{2}\|p - q\|. \tag{1.5}$$

Hence Ψ is a contraction and $\|\Psi(p)\| \le \frac{1}{2}\|p\|$ for all $\|p\| < r$. For $v \in X$ we set

$$\Phi_v(u) = v - \Psi(u).$$

Trivially, Φ_v is a contraction. Moreover, for all $u \in B(r)$ and $v \in B(r/2)$ there results

$$\|\Phi_v(u)\| \le \|v\| + \|\Psi(u)\| \le r.$$

Hence, for $\|v\| \le r/2, \Phi_v$ is a contraction, maps $B(r)$ into itself and therefore has a unique fixed point $u \in B(r)$, which satisfies

$$u = \Phi_v(u) = v - \Psi(u),$$

namely $F(u) = v$. As a consequence, we can define the inverse $F^{-1} : B(r/2) \to B(r)$. To show that F^{-1} is continuous, we let $u = F^{-1}(v)$ and $w = F^{-1}(z)$, that is,

$$\left.\begin{array}{l} u + \Psi(u) = v, \\ w + \Psi(w) = z. \end{array}\right\}$$

Using (1.5) we get that

$$\|u - w\| \leq \|v - z\| + \|\Psi(u) - \Psi(w)\| \leq \|v - z\| + \frac{1}{2}\|u - w\|$$

and thus

$$\|F^{-1}(v) - F^{-1}(z)\| \leq 2\|v - z\|.$$

This proves that F^{-1} is continuous, indeed Lipschitz-continuous with constant 2. In particular, letting $V = B(r/2), U = B(r) \cap F^{-1}(V)$, we get

$$F|_U \in \mathrm{Hom}(U, V).$$

(ii) Setting $u = F^{-1}(v)$, from $u + \Psi(u) = v$, we get

$$F^{-1}(v) = v - \Psi(F^{-1}(v)).$$

Since $\Psi(u) = o(\|u\|)$ and recalling that F^{-1} is Lipschitz-continuous, we infer that $\Psi(F^{-1}(v)) = o(\|v\|)$. This shows that F^{-1} is differentiable at $v = 0$ and

$$\mathrm{d}F^{-1}(0) = I.$$

Then, if $v \in B(r/2)$ and $u = F^{-1}(v)$, after a translation that carries u and v into the origins of X and Y, respectively, one can infer that F^{-1} is differentiable at v and

$$\mathrm{d}F^{-1}(v) = (F'(u))^{-1}.$$

This proves that (1.3) holds.

Next, we remark that the map $(F^{-1})' : v \to (F'(F^{-1}(v)))^{-1}$ is obtained by composition in the following way:

$$v \xrightarrow{F^{-1}} u = F^{-1}(v) \xrightarrow{F'} F'(u) \xrightarrow{J} (F'(u))^{-1} = J[F'(u)]$$

Since all the maps F^{-1}, F' and J are continuous, it follows that $F^{-1} \in C^1$, proving (ii).

(iii) Let $F \in C^k$. By induction on k, suppose that $F^{-1} \in C^{k-1}$. Repeating the above arguments and taking into account that $J \in C^\infty$ (Proposition 1.1) we get readily that $F^{-1} \in C^k$.

Remark 1.3 The regularity assumption $F \in C^1$ cannot be eliminated, in general. If X and Y are finite-dimensional, elementary examples show that we can drop injectivity. To see this, let $\varphi : \mathbb{R} \to \mathbb{R}$ be non-decreasing and such that

$$\varphi(s) = \begin{cases} \dfrac{1}{n} & \text{for} & \dfrac{1}{n} - \dfrac{1}{4n^2} \leq s \leq \dfrac{1}{n} + \dfrac{1}{4n^2}, \\ s + O(s^2) & \text{as} & s \to 0. \end{cases}$$

Plainly, φ is differentiable at $s = 0$ (indeed, it can be chosen to be C^∞

in $\mathbb{R} - \{0\}$) and there results $\varphi'(0) = 1$, but φ is not injective in any neighbourhood of $s = 0$.

In the case of infinite-dimensional spaces, the same φ provides a counter-example for surjectivity, too.

Let $X = Y = C([-1, 1])$ and $F : X \to Y$ be given by $F(u) = \varphi \circ u$. Consider the sequence $v_n \in Y$,

$$v_n(t) = \frac{1}{n} + \frac{t}{n^2}.$$

For such a sequence there result $\|v_n\| \to 0$ and $v_n \notin R(F)$. In fact if there exist $u_n \in X$ such that $F(u_n) = v_n$ one would find

$$\varphi(u_n(t)) = v_n(t) = \frac{1}{n} + \frac{t}{n^2}.$$

Hence there results

$$\varphi(u_n(t)) > \frac{1}{n} \text{ for } t > 0,$$

while

$$\varphi(u_n(t)) < \frac{1}{n} \text{ for } t < 0.$$

Then the monotonicity of φ implies

$$u_n(t) \geq \frac{1}{n} + \frac{1}{4n^2} \text{ for } t > 0,$$

$$u_n(t) \leq \frac{1}{n} - \frac{1}{4n^2} \text{ for } t < 0,$$

and u_n is not continuous at $t = 0$, i.e. $u_n \notin X$.

Note that F is differentiable at $u = 0$ with derivative $F'(0) = I_X$, but $F \notin C^1(X, Y)$.

The Local Inversion Theorem is the rigorous justification of the so-called "procedure by linearization". Roughly, it permits us to solve, locally, a nonlinear problem through the study of its linearization.

Some easy examples will illustrate the main steps of the procedure. Further applications are postponed to Section 3 below.

Example 1.4   Let us seek the T-periodic solutions $x = x(t)$ of

$$x'' + g(x, x') = \varepsilon h(t) \tag{1.6}$$

where $g \in C^1(\mathbb{R} \times \mathbb{R}, \mathbb{R})$ and $h \in C(\mathbb{R})$ is T-periodic.

In accordance with (1.6) we set

$$X = \{x \in C^2(\mathbb{R}, \mathbb{R}) | x(t + T) = x(t) \text{ for all } t \in \mathbb{R}\},$$

$$Y = \{h \in C(\mathbb{R}, \mathbb{R}) | h(t + T) = h(t) \text{ for all } t \in \mathbb{R}\},$$

and

$$F(x) = x'' + g(x, x').$$

Suppose $g(0,0) = 0$ in such a way that (1.6) has for $\varepsilon = 0$ the "trivial" solution $x = 0$ and we shall apply Theorem 1.2 with $u^* = 0$

It is immediately verifiable that $F \in C^1(X, Y)$ and $F'(0)[w] = w'' + aw' + bw$, where $a = g_x, (0,0)$ and $b = g_x(0,0)$.

As an immediate consequence of the Fredholm Alternative Theorem, $F'(0)$ is invertible whenever the linear equation $w'' + aw' + bw = 0$, $w \in X$, has only the trivial solution $w = 0$. If this is the case, then there are ε^* and $\delta > 0$ such that for all $|\varepsilon| < \varepsilon^*$ (1.6) has a unique solution x with $\|x\|_X < \delta$.

Example 1.5 Let Ω be a bounded domain in $\mathbb{R}^n$ with smooth boundary $\partial\Omega$. Consider the boundary-value problem

$$\left.\begin{array}{l} \Delta u + \lambda u - u^3 = h(x), \text{ in } \Omega; \\ u = 0, \text{ in } \partial\Omega \end{array}\right\} \tag{1.7}$$

where Δ is the Laplace operator and $\lambda \in \mathbb{R}$.

Here we let $X = \{u \in C^{2,\alpha}(\overline{\Omega}) : u(x) = 0 \text{ on } \partial\Omega\}$, $Y = C^{0,\alpha}(\overline{\Omega})$ and associate to (1.7) the map $F : X \to Y$ defined by

$$F(u) = \Delta u + \lambda u - u^3.$$

Problem (1.7) leads us to seek, for any given $h \in Y$, a $u \in X$ such that $F(u) = h$. Let us apply Theorem 1.2 with $u^* = 0$.

Plainly f is $C^\infty(X, Y)$ and one has $F'(u)[w] = \Delta w + \lambda w - 3u^2 w$.

Hence for $\lambda \neq \lambda_k$, the eigenvalues of the Laplace operator Δ on Ω (with zero boundary conditions), the linear problem

$$F'(0)[w] = \Delta w + \lambda w = 0, \ w \in X,$$

has only the solution $u = 0$. By the Fredholm Alternative it follows that, whenever $\lambda \neq \lambda_k, F'(0)$ is one-to-one from X onto Y. As a consequence of the Closed Graph Theorem $[F'(0)]^{-1}$ (exists and) is continuous and we can conclude that if $\lambda \neq \lambda_k$ then for all $h \in Y$ with norm $\|h\|_Y$ small enough, (1.7) has a unique solution $u = u(h) \in X$ with norm $\|u\|_X$ small. Moreover the correspondence $h \to u(h)$ is C^∞.

Example 1.6 Let Ω be a bounded connected domain in $\mathbb{R}^2$, with smooth boundary $\partial\Omega$, and let γ be a smooth function on $\partial\Omega$. A smooth function $u : \Omega \to \mathbb{R}$ is a minimal surface with boundary γ if

$$\left.\begin{array}{l} \mathcal{M}(u) = Au_{xx} + Bu_{yy} - 2u_x u_y u_{xy} = 0, \\ u|_{\partial\Omega} = \gamma, \end{array}\right\} \tag{1.8}$$

where $A = (1 + u_y^2)$ and $B = (1 + u_x^2)$.

We shall work in Hölder spaces (see Subsection 0.5). Note that, if $\partial\Omega$

is of class $C^{m,\alpha}$ that is $\partial\Omega$ is locally $C^{m,\alpha}$-diffeomorphic to a segment, we can consider the space $C^{m,\alpha}(\partial\Omega)$.

Let $X = C^{2,\alpha}(\overline{\Omega})$, $Y = C^{0,\alpha}(\overline{\Omega}) \times C^{2,\alpha}(\partial\Omega)$ and $F : X \to Y$ be given by

$$F(u) = (\mathcal{M}(u), u|_{\partial\Omega}).$$

We are now ready to apply the Local Inversion Theorem at $u^* = 0$. It is easy to check that $F \in C^1(X, Y)$ and

$$F'(u)[w] = (Aw_{xx} + Bw_{yy} - 2u_x u_y w_{xy} + 2u_y u_{xx} w_y$$
$$+ 2u_x u_{yy} w_x + 2(u_x w_y + u_y w_x)u_{xy}, w|_{\partial\Omega}).$$

For $u = 0$ it follows that

$$F'(0)[w] = (\Delta w, w|_{\partial\Omega}).$$

According to Theorem 0.8, the Dirichlet problem

$$\left.\begin{array}{l} \Delta w = h \ \text{ in } \Omega, \\ w = \phi \ \text{ on } \partial\Omega, \end{array}\right\}$$

has a unique solution $w \in X$, provided $(h, \phi) \in Y$, and w depends continuously upon the data h, ϕ.

Then Theorem 1.2 yields the existence of neighbourhoods U and V of 0 in X and $C^{2,\alpha}(\partial\Omega)$, respectively, such that if $\gamma \in V$ then (1.8) has a unique solution $u \in U$, and the correspondence $\gamma \to u$ is C^1.

It is worth noticing that (1.8) can have no solutions for some γ if Ω is not convex (see [**CH**, Vol.II, p.167]).

2 The Implicit Function Theorem

Often the introduction of a parameter permits us to extend the range of applicability of the Local Inversion Theorem.

Let us consider maps $F : \Lambda \times U \to Y$, where Λ and U are open subsets of Banach spaces T and X, respectively, and Y is a Banach space.

We start with the following lemma.

Lemma 2.1 *Let* $(\lambda^*, u^*) \in \Lambda \times U$. *Suppose that*

(i) *F is continuous and F has the u-partial derivative in $\Lambda \times U$ and $F_u : \Lambda \times U \to L(X, Y)$ is continuous.*

(ii) *$F_u(\lambda^*, u^*) \in L(X, Y)$ is invertible.*

Then the map $\Psi : \Lambda \times U \to T \times Y$, given by

$$\Psi(\lambda, u) = (\lambda, F(\lambda, u)), \tag{2.1}$$

is locally invertible at (λ^*, u^*) with continuous inverse Φ.

If, in addition, $F \in C^1(\Lambda \times U, Y)$ then Φ is C^1.

Proof. To prove that Ψ is locally invertible at (λ^*, u^*) it suffices to repeat, with obvious changes, the proof of Theorem 1.2.

Next, suppose that $F \in C^1(\Lambda \times U, Y)$ and let

$$A = F_\lambda(\lambda^*, u^*), \quad B = F_u(\lambda^*, u^*),$$

Plainly $\Psi \in C^1(\Lambda \times U, T \times Y)$ with derivative

$$\Psi'(\lambda^*, u^*)(\xi, v) = (\xi, A[\xi] + B[v]).$$

The equation

$$\Psi'(\lambda^*, u^*)(\xi, v) = (\eta, v)$$

yields $\xi = \eta$, and

$$A[\eta] + B[v] = v, \tag{2.2}$$

Since B is invertible (assumption (ii)) (2.2) has a unique solution $v = B^{-1}(v - A[\eta])$. It follows that $\Psi'(\lambda^*, u^*) \in \mathrm{Inv}(T \times Y, T \times Y)$. An application of the Local Inversion Theorem to Ψ proves that (Ψ is locally invertible at (λ^*, u^*) and) the inverse Φ is C^1.

Remarks 2.2

(i) Under the assumptions of Lemma 2.1 Ψ has an inverse Φ defined in a neighbourhood $\Theta \times V$ of $(\lambda^*, F(\lambda^*, u^*))$. Owing to the definition (2.1) of Ψ, the first component of Φ is the identity. In other words there results

$$\Phi(\lambda, v) = (\lambda, \varphi(\lambda, v)) \tag{2.3}$$

for some $\varphi : \Theta \times V \to X$ satisfying

$$F(\lambda, \varphi(\lambda, v)) = v \quad \text{for all } \lambda \in \Theta. \tag{2.4}$$

Such a φ is of class C^1 and its derivatives φ_λ and φ_v can be found by differentiating the identity (2.4):

$$F_\lambda + F_u \circ \varphi_\lambda = 0,$$

$$F_u \circ \varphi_v = I.$$

It follows that

$$\varphi_\lambda = -[F_u]^{-1} F_\lambda, \tag{2.5}$$

and

$$\varphi_v = [F_u]^{-1}.$$

(ii) The existence of the local inverse Φ of Ψ can be proved by taking F defined on $\Lambda \times U$, Λ being an open subset of a *topological space T*.

We are now in a position to state the Implicit Function Theorem.

Theorem 2.3 (Implicit Function Theorem)   *Let* $F \in C^k(\Lambda \times U, Y)$, $k \geq 1$, *where* Y *is a Banach space and* Λ *(resp. U) is an open subset of Banach space* T *(resp. X). Suppose that* $F(\lambda^*, u^*) = 0$ *and that* $F_u(\lambda^*, u^*) \in \mathrm{Inv}(X, Y)$.

Then there exist neighbourhoods Θ *of* λ^* *in* T *and* U^* *of* u^* *in* X *and a map* $g \in C^k(\Theta, X)$ *such that*

(i) $F(\lambda, g(\lambda)) = 0$ *for all* $\lambda \in \Theta$,

(ii) $F(\lambda, u) = 0, (\lambda, u) \in \Theta \times U^*$, *implies* $u = g(\lambda)$,

(iii) $g'(\lambda) = -[F_u(p)]^{-1} \circ F_\lambda(p)$, *where* $p = (\lambda, g(\lambda))$ *and* $\lambda \in \Theta$.

Proof.   First we associate to F the map Ψ given by (2.1). According to Lemma 2.1, Ψ is locally invertible at (λ^*, u^*) and $\Psi(\lambda^*, u^*) = (\lambda^*, F(\lambda^*, u^*)) = (\lambda^*, 0)$.

Using Remark 2.2 (i), we find that the inverse Φ of Ψ satisfies (2.3). It is also easy to verify that $\varphi \in C^k$ provided F is C^k.

Setting

$$g(\lambda) = \varphi(\lambda, 0) \quad (\lambda \in \Theta),$$

and using (2.4) we get

$$F(\lambda, g(\lambda)) = F(\lambda, \varphi(\lambda, 0)) = 0, \quad \text{for all } \lambda \in \Theta,$$

which proves (i).

Since Φ is one-to-one, (ii) follows, too. Lastly, (iii) is an immediate consequence of (2.5).

Example 2.4   If $T = X = Y = \mathbb{R}$ the above theorem yields a (unique) C^1 Cartesian curve $u = g(\lambda)$, defined in a neighbourhood of $\lambda = \lambda^*$, such that $F(\lambda, g(\lambda)) = 0$, which is nothing else than the elementary Implicit Function Theorem.

3 A stability property of orbits

Many perturbation problems in analysis can be handled by the abstract results of sections 1 and 2.

Here we discuss in detail an important application to the existence of periodic solutions of perturbed differential systems.

Non-autonomous systems

Given $f \in C^1(\mathbb{R} \times \mathbb{R} \times \mathbb{R}^n, \mathbb{R}^n)$ let us consider the system of ordinary differential equations

$$x' =: \frac{\mathrm{d}x}{\mathrm{d}t} = f(\varepsilon, t, x). \qquad (3.1.\varepsilon)$$

We suppose that

$$f(\varepsilon, t + T, x) = f(\varepsilon, t, x) \quad \text{for all } (\varepsilon, t, x) \in \mathbb{R} \times \mathbb{R} \times \mathbb{R}^n, \qquad (3.2)$$

and

$$\text{for } \varepsilon = 0 \; (3.1.0) \; \text{has a } T - \text{periodic solution } y = y(t). \qquad (3.3)$$

We want to discuss the situation if $(3.1.\varepsilon)$ has, for ε small, a T-periodic solution y_ε close to y.

Let us consider the Cauchy problem

$$\left. \begin{array}{l} \alpha' = f(\varepsilon, t, \alpha), \\ \alpha(0) = \xi. \end{array} \right\} \qquad (3.4)$$

Since f is C^1, (3.4) has a unique solution $\alpha = \alpha(\varepsilon, t, \xi)$ defined for $|\varepsilon|$ small, ξ in a neighbourhood of $\xi^* = y(0)$ and $t \in [0, T]$. Moreover, it is well known that α is differentiable with respect to the initial value ξ and the derivative

$$A(\varepsilon, t, \xi) := \frac{\partial \alpha}{\partial \xi}$$

is the $n \times n$ matrix solving the Cauchy problem

$$\left. \begin{array}{l} \dfrac{\mathrm{d}A}{\mathrm{d}t} = f_x(\varepsilon, t, \alpha)A, \\[2mm] A(\varepsilon, 0, \xi) = I = I_{\mathbb{R}^n}. \end{array} \right\}$$

For $\varepsilon = 0$ and $\xi = \xi^*$ the matrix $A(0, t, \xi^*)$ will be denoted by $A_0(t)$.

Theorem 3.1 *Suppose (3.2), (3.3) hold and*

$$\lambda = 1 \text{ is not in the spectrum } \sigma \text{ of } A_0(T). \qquad (3.5)$$

Then there exists $\delta > 0$ and a continuous $\xi(\varepsilon)$, $|\varepsilon| < \delta$, with $\xi(0) = \xi^*$, such that for $|\varepsilon| < \delta$ $(3.1.\varepsilon)$ possesses a unique T-periodic solution y_ε with $y_\varepsilon(0) = \xi(\varepsilon)$.

Proof. It is evident that $(3.1.\varepsilon)$ has a T-periodic solution if and only if there exists $\xi \in \mathbb{R}^n$ such that

$$\alpha(\varepsilon, T, \xi) = \xi.$$

Hence, introducing the map $F : \mathbb{R} \times \mathbb{R}^n \to \mathbb{R}^n$, defined by

$$F(\varepsilon, \xi) = \alpha(\varepsilon, T, \xi) - \xi,$$

we are led to solve the equation $F(\varepsilon, \xi) = 0$.

Notice that $F \in C^1(\mathbb{R} \times \mathbb{R}^n, \mathbb{R}^n)$. Moreover, since $\alpha(0, t, \xi^*) = y(t)$ and y is T-periodic, it follows that

$$F(0, \xi^*) = \alpha(0, T, \xi^*) - \xi^* = y(T) - \xi^* = 0.$$

In order to apply the Implicit Function Theorem, we evaluate

$$F_\xi(0, \xi^*) = \alpha_\xi(0, T, \xi^*) - I = A_0(T) - I.$$

By (3.5) $\lambda = 1 \notin \sigma$ and therefore $F_\xi(0, \xi^*)$ is invertible. Hence for $|\varepsilon|$ small $F(\varepsilon, \xi) = 0$ has a unique solution $\xi = \xi(\varepsilon)$ near $\xi = \xi^*$, proving the theorem.

Autonomous systems

The case in which f is independent of time is more delicate and requires additional study.

In this case system $(3.2.\varepsilon)$ becomes

$$x' = f(\varepsilon, x). \qquad (3.6.\varepsilon)$$

Let us point out that the period of possible solutions of $(3.6.\varepsilon)$ is *a priori* unknown. This makes the problem more difficult to solve and often the perturbation results are the only ones that can be achieved in this case. We will set $f(0, x) = f(x)$.

Let us assume $f \in C^1(\mathbb{R}^n, \mathbb{R}^n)$ and that

the unperturbed system $x' = f(x)$ has a non $-$ constant

$$T-periodic \ solution \ y = y(t). \qquad (3.7)$$

Without loss of generality we can suppose that $y(0) = 0$. Let us remark explicitly that, since y is not constant, $y'(t) \neq 0$ *for all t*.

Remark 3.2 Theorem 3.1 does not apply in the case of autonomous systems like $(3.6.\varepsilon)$. In fact here $A_0 = A(0, t, 0)$ satisfies

$$\left. \begin{array}{l} \dfrac{\mathrm{d}A_0}{\mathrm{d}t} = f'(y(t))A_0, \\[2mm] A_0(0) = I, \end{array} \right\} \qquad (3.8)$$

and we claim that $\lambda = 1 \in \sigma$, the spectrum of the matrix $A_0(T)$.

To see this, we differentiate

$$y' = f(y)$$

and find

$$y'' = f'(y)y'.$$

Hence, setting $v = y'$ we get $v(t) \neq 0$ for all t and

$$v' = f'(y)v. \qquad (3.9)$$

Let $v^* = v(0)$ and $w(t) = A_0(t)v^*$. From (3.8) it follows that

$$\left.\begin{aligned} w' &= \frac{\mathrm{d}A_0}{\mathrm{d}t}v^* = f'(y)A_0(t)v^* = f'(y)w, \\ w(0) &= v^*. \end{aligned}\right\} \qquad (3.10)$$

By the uniqueness of the Cauchy problem, (3.9) and (3.10) yield $w(t) = v(t)$. In particular, there results $w(T) = w(0)$ and hence

$$A_0(T)v^* = w(0) = v^*.$$

Since $v^* \neq 0$, $\lambda = 1$ is an eigenvalue of $A_0(T)$, proving our claim.

According to the above remark, $v^* \in \mathrm{Ker}(A_0(T) - I)$. We shall suppose that

$$\lambda = 1 \text{ *is a simple eigenvalue of* } A_0(T) \qquad (3.11)$$

that is (see §0.4), that, letting $M = A_0(T) - I$ and $W = R(M)$, we get

$$\mathrm{dim}(\mathrm{Ker}(M)) = \mathrm{codim}\,(W) = 1, \qquad (3.11\text{-i})$$

$$\mathrm{Ker}(M) = \mathrm{Ker}(M^2). \qquad (3.11\text{-ii})$$

Theorem 3.3 *Suppose that* (3.7) *and* (3.11) *hold. Then for* $|\varepsilon|$ *small there exist continuous mappings* $h = h(\varepsilon)$ *and* $\tau = \tau(\varepsilon)$ *such that*

$$h(0) = y(0), \quad \tau(0) = T,$$

and (3.6.ε) *has a* $\tau(\varepsilon)$-*periodic solution* y_ε *satisfying* $y_\varepsilon(0) = h(\varepsilon)$.

Remark 3.4 From the geometrical point of view Theorem 3.3 ensures that the orbit $\{y_\varepsilon(t)\}_{t \in \mathbb{R}}$ is close to the orbit $\{y(t)\}_{t \in \mathbb{R}}$. Moreover this solution is the only one having period close to T and orbit close to $\{y(t)\}_{t \in \mathbb{R}}$.

Proof of Theorem 3.3 Without loss of generality we can assume $|v^*| = 1$. Let us consider the hyperplane

$$\mathcal{H} = \{h \in \mathbb{R}^n : h \cdot v^* = 0\} \quad (\cong \mathbb{R}^{n-1}).$$

Roughly, the solution $y(t)$ leaves $\mathcal{H}$ at the time $t = 0$ from the point $y(0) = 0$ and comes back to the same point $y(T) = 0$ at the time $t = T$. For $h \in \mathcal{H}$ near 0 the solution $\alpha_\varepsilon = \alpha(\varepsilon, t, h)$ of

$$\left.\begin{aligned} \frac{\mathrm{d}\alpha_\varepsilon}{\mathrm{d}t} &= f(\varepsilon, \alpha), \\ \alpha(\varepsilon, 0, h) &= h, \end{aligned}\right\}$$

will reach $\mathcal{H}$ at a certain time τ close to T. (See Fig. 2.1)

Our goal will be to show that there exist $h \in \mathcal{H}$ and $\tau \in \mathbb{R}$ such that $\alpha(\varepsilon, \tau, h) = h$ and this will provide a τ-periodic solution of (3.6.ε).

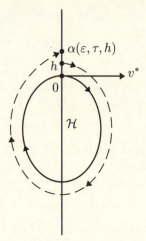

Figure 2.1

Set $u = (t, p) \in \mathbb{R} \times \mathbb{R}^n = X$ and let

$$F : \mathbb{R} \times X \to X$$

be the "*Poincaré map*" defined by

$$F : (\varepsilon, t, p) \to (p \cdot v^*, \alpha(\varepsilon, t, p) - p) \in X.$$

If $F(\varepsilon, \tau, p) = (0, 0)$ for some $(\varepsilon, \tau, p) \in \mathbb{R} \times X$, then

$$p \cdot v^* = 0$$

and

$$\alpha(\varepsilon, \tau, p) = p.$$

According to the preceding discussion this means that $(3.6.\varepsilon)$ has a τ-periodic solution y_ε and $y_\varepsilon(0) = p \in \mathcal{H}$.

We are going to apply the Implicit Function Theorem to F, taking ε as parameter. There results $F(0, T, 0) = (0, \alpha(0, T, 0)) = (0, 0)$. Moreover $F \in C^1(\mathbb{R} \times X, X)$ and the derivative $F_u(\varepsilon, u) \in L(X, X)$ is given by

$$F_u(\varepsilon, u) : (a, v) \to (v \cdot v^*, \alpha_t(\varepsilon, t, p)[a] + \alpha_\xi(\varepsilon, t, p)[v] - v).$$

In order to evaluate the second component of $F_u(0, T, 0)$ we start by recalling that $\alpha_0(t) = \alpha(0, t, 0)$ satisfies $d\alpha_0/dt = F(\alpha_0), \alpha_0(0) = 0$. Thus $\alpha_0(t) = y(t)$ and, in particular, one has

$$\alpha_t(0, T, 0)[a] = ay'(T) = ay'(0) = av^*.$$

As for $\alpha_\xi(0, T, 0)$, this is nothing but the matrix $A_0(T)$. In conclusion there results

$$F_u(0, T, 0)(a, v) = (v \cdot v^*, av^* + A_0(T)[v] - v)$$
$$= (v \cdot v^*, av^* + M(v)).$$

Next, we need a lemma.

Lemma 3.5 $F_u(0, T, 0) \in \mathrm{Inv}(X, X)$.

Proof. Given $(b, z) \in X(= \mathbb{R} \times \mathbb{R}^n)$, we have to solve the system

$$\left. \begin{array}{l} v \cdot v^* = b, \\ av^* + M(v) = z. \end{array} \right\}$$

By assumption (3.11 (i)) there exists $z^* \in \mathbb{R}^n$, $|z^*| = 1$, such that $\mathbb{R}^n = \mathbb{R}z^* \oplus W$, with $z^* \cdot w = 0$ for all $w \in W$, and hence any $z \in \mathbb{R}^n$ can be written in the form $z = sz^* + w$, with $s \in \mathbb{R}, w \in W$. Moreover each $v \in \mathbb{R}^n$ can be written as $v = rv^* + h$, with $r \in \mathbb{R}$ and $h \in \mathcal{H}$.

Substituting into the preceding system and taking into account that $h \cdot v^* = 0$ and $M(v^*) = 0$, we find

$$r = b$$

and

$$av^* + M(h) = sz^* + w, \tag{3.12}$$

Taking the projections of (3.12) onto $\mathbb{R}z^*$ and W, respectively, and recalling that $M(h) \in W$, we find that (3.12) becomes

$$a(v^* \cdot z^*) = s, \tag{3.13}$$

$$av^* - a(v^* \cdot z^*)z^* + M(h) = w. \tag{3.14}$$

If $v^* \in W$, there exists $q \in \mathbb{R}^n$ $(q \neq 0)$ such that $Mq = v^*$ and hence $M^2 q = Mv^* = 0$. Then $q \in \mathrm{Ker}(M^2)$, whereas $q \notin \mathrm{Ker}(M)$. This means that $\mathrm{Ker}(M) \subsetneq \mathrm{Ker}(M^2)$, contradicting (3.11(ii)). Therefore $v^* \cdot z^* \neq 0$, and (3.13) yields

$$a = \frac{s}{v^* \cdot z^*}. \tag{3.15}$$

Substituting into (3.14) one finds

$$M(h) = w - \frac{s}{v^* \cdot z^*} v^* + sz^* \in W. \tag{3.16}$$

Since M is invertible from $\mathcal{H}$ to W, (3.16) has a unique solution $h = \varphi(s, w)$ with φ continuous.

Then the solution of equation $F_u(0, T, 0)[a, v] = (b, z) = (b, sz^* + w)$ is given by $a = (3.15)$ and $v = bv^* + \varphi(s, w)$ and the claim follows.

Proof of Theorem 3.3 completed. Lemma 3.5 enables us to apply the Implicit Function Theorem to F. Then there exist $\tau(\varepsilon)$ and $h(\varepsilon)$, defined in a suitable neighbourhood of $\varepsilon = 0$, such that $F(\varepsilon, \tau(\varepsilon), h(\varepsilon)) = (0, 0)$.

In particular, $h(\varepsilon) \in \mathcal{H}$ and

$$\alpha(\varepsilon, \tau(\varepsilon), h(\varepsilon)) = h(\varepsilon).$$

Hence $y_\varepsilon(t) := \alpha(\varepsilon, t, h(\varepsilon))$ gives rise to a $\tau(\varepsilon)$-periodic solution of $(3.6.\varepsilon)$ such that $y_\varepsilon(0) = h(\varepsilon)$.

3

Global inversion theorems

This chapter deals with the extension in the large of the local results discussed above.

Section 1 contains the Global Inversion Theorem (sometimes called "Monodromy Theorem"), which goes back to Hadamard in the finite-dimensional case, and to Caccioppoli [**Ca**] and P. Levy [**Le**] for general Banach spaces. Even if such a theorem is a classical result often understood in the current literature we think useful to have given here an elementary version, in the frame of Banach spaces.

In Section 2 we deal with mappings F that possess singularities and are not global homeomorphisms. Following [**AP**], we study the case of singularities corresponding to a one-dimensional kernel of F', when a complete, geometric description of the range can be given.

1 The Global Inversion Theorem

In this section we want to investigate conditions under which a map F is a global homeomorphism. Since the results we are going to state are topological in nature, we will consider a map $F : M \to N$, where M and N denote metric spaces.

Let $F : M \to N$ and for any subset A of N let

$$F^{-1}(A) = \{u \in M : F(u) \in A\}$$

denote the pre-image of A through F. For brevity, we will write $F^{-1}(u)$ for $F^{-1}(\{u\})$.

We need the following.

Definition 1.1 We say that F is *proper* if $F^{-1}(K)$ is compact (in M) for all compact set $K \subset N$.

Let us remark that if F is proper then it maps closed sets into closed sets.

For all $v \in N$ we let $[v]$ denote the *cardinal number* of the set $F^{-1}(v)$

Theorem 1.2 *Suppose $F \in C(M, N)$ is proper and locally invertible in M. Then $[v]$ is finite for all $v \in N$ and locally constant.*

Proof. For all $v \in N, F^{-1}(v)$ is compact (since F is proper) and discrete (since F is locally invertible); hence $[v]$ is finite.

Next, fixing an arbitrary $v \in N$, let

$$F^{-1}(v) = \{u_1, u_2, \ldots, u_k\}.$$

Since F is locally invertible at each u_i ($i = 1, 2, \ldots, k$) we can find neighbourhoods U_i of u_i and V of v such that $F|_{U_i} \in \mathrm{Hom}(U_i, V)$. Notice that

$$[q] \geq k \text{ for all } q \in V, \tag{1.2}$$

because the equation $F(u) = v$ has at least k solutions, one on each U_i.

We claim that there is a neighbourhood $W \subset V$ of v such that

$$[w] = k \text{ for all } w \in W. \tag{1.3}$$

If not, there would be a sequence $v_n \to v$, $v_n \in V$, such that $[v_n] \neq k$. According to (1.2) we can infer that $[v_n] > k$ and hence there exist points $p_n \notin \cup_{1 \leq i < k} U_i$, with $F(p_n) = v_n$.

Since F is proper, up to a subsequence, $p_n \to p$. By continuity $F(p) = v$, and therefore p belongs to some U_j, a contradiction because $p_n \notin U_j$ and $p_n \to p$. This proves (1.3) and completes the proof of the theorem.

Remark 1.3 If F is not proper, $[v]$ could be infinite and not locally constant. For example, this is the case if $M = N = \mathbf{C}$ and $F(z) = \exp(z)$. Notice that F is locally invertible at each point $z \in \mathbf{C}$.

As an immediate consequence of Theorem 1.2 one has the following.

Corollary 1.4 *Suppose $F \in C(M, N)$ is proper and locally invertible in M, and let N be connected. Then $[v]$ is constant for all $v \in N$.*

In order to improve Theorem 1.2 we give the following definition.

Definition 1.5 A *singular* point is a $u \in M$ where F is not locally invertible. The set of all singular points of F will be denoted by Σ. We will also set

$$\Sigma_0 = F^{-1}(F(\Sigma)),$$
$$M_0 = M\backslash\Sigma_0,$$
$$N_0 = N\backslash F(\Sigma).$$

It is worth noticing that since Σ is closed and F is proper $F(\Sigma)$ is closed in N (see the remark after Definition 1.1). Thus Σ_0 is closed and M_0 and N_0 are open.

Theorem 1.6 *Let $F \in C(M,N)$ be proper. Then $[v]$ is locally constant on every connected component of $N\backslash F(\Sigma)$.*

Proof. Let us consider the restriction $F^* = F|_{M_0}$

$$F^* : M_0 \to N_0.$$

Obviously, F^*, as a map from M_0 to N_0, is locally invertible at any $u \in M_0$ and is proper. Then, applying Corollary 1.4 with $M = M_0$ and $N = N_0$, we get the result.

The preceding result can be greatly improved if N_0 is *simply connected*.

Recall that a topological space T is simply connected if it is arcwise connected and every closed path σ in T is homotopic to a constant. In other words, given any

$$\sigma \in C([0,1],T), \text{ with } \sigma(0) = \sigma(1),$$

there exist $h \in C([0,1] \times [0,1],T)$ and $v \in T$ such that

$$\left.\begin{array}{l} h(s,0) = \sigma(s) \text{ for all } s, \\ h(s,1) = v \text{ for all } s, \\ h(0,t) = h(1,t) \text{ for all } t. \end{array}\right\} \tag{1.4}$$

Theorem 1.7 *Suppose $F \in C(M,N)$ is proper and let $N_0 = N\backslash F(\Sigma)$ be simply connected and $M_0 = M\backslash F^{-1}(F(\Sigma))$ be arcwise connected. Then F is a homeomorphism from M_0 onto N_0.*

As a corollary, when Σ is empty, we obtain a very classical result.

Global Inversion Theorem 1.8 *Let $F \in C(M,N)$ be proper and locally invertible on all of M. Suppose that M is arcwise connected and N is simply connected. Then F is a homeomorphism from M onto N.*

The proof of Theorem 1.7 will be carried out through several steps. It is convenient to introduce a definition. Let M, N be metric spaces and let $F \in C(M, N)$.

Definition 1.9 Given a path $\sigma : [a, b] \to N$; we say that the path $\theta : [a, b] \to M$ *inverts* F *along* σ if $\sigma = F \circ \theta$; namely if the diagram

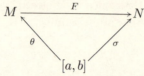

commutes. In such a case, we will also say that F is *invertible along* σ, *with inverse* θ.

Remarks 1.10 (i) Let $u \in M$ and $v \in N$ be such that $F(u) = v$ and suppose that there are neighbourhoods U and V of u and v, respectively, such that $F|_U \in \mathrm{Hom}(U, V)$. Given any path $\sigma : [a, b] \to N$ with $\sigma(a) = v$ and such that $\sigma[a, b] \subset V$, then the relationship $F(\theta(t)) = \sigma(t)$ defines a path θ which inverts F along σ, and such that $\theta(a) = u$. Moreover θ is the only path with the above properties.

(ii) (Construction of inverting paths by "pasting") Let $\sigma : [a, b] \to N$ be a path and let $c \in (a, b)$ be given. Suppose that there are two paths θ_1 and θ_2 such that $\theta_1 : [a, c] \to M$ inverts F along $\sigma|_{[a,c]}$, $\theta_2 : [c, b] \to M$ inverts F along $\sigma|_{[c,b]}$. Moreover let $\theta_1(c) = \theta_2(c)$.

Then, defining $\theta : [a, b] \to M$ by setting $\theta|_{[a,c]} = \theta_1$ and $\theta|_{[c,b]} = \theta_2$, we have that θ inverts F along σ. Indeed, θ is well defined and is continuous at the point c.

Lemma 1.11 *Let* $u^* \in M_0$ *and* $v^* = F(u^*) \in N_0$. *Then given any path* $\sigma : [0, 1] \to N_0$ *such that* $\sigma(0) = v^*$, *there exists a unique path* $\theta : [0, 1] \to M_0$ *that inverts* F *along* σ *such that* $\theta(0) = u^*$.

Proof. (Uniqueness). Let θ_1 and θ_2 be two paths which invert F along σ with $\theta_1(0) = \theta_2(0) = u^*$ and let

$$\xi = \sup\{s \in [0, 1] : \theta_1(t) = \theta_2(t), \text{ in } [0, s]\}.$$

According to Remark 1.10 (i) ξ is well defined and $\xi > 0$ because $u^* \in M_0$. By continuity one has plainly that $\theta_1(\xi) = \theta_2(\xi)$. Let us suppose the contrary that $\xi < 1$ and set

$$u = \theta_1(\xi) = \theta_2(\xi), \quad v = F(u).$$

Since F is locally invertible on M_0, there are neighbourhoods U of u and V of v such that $F|_U \in \mathrm{Hom}(U, V)$.

Moreover, since θ_1 and θ_2 are continuous, $\exists\ \alpha > 0$ such that

$$\theta_1([\xi, \xi + \alpha]) \subset U, \quad \theta_2([\xi, \xi + \alpha]) \subset U.$$

On the other side, since $F(\theta_1(t)) = F(\theta_2(t)) = \sigma(t)$, $\theta_1(t) = \theta_2(t)$ for $t \in [\xi, \xi + \alpha]$. It follows that θ_1 and θ_2 coincide on $[0, \xi + \alpha]$, in contradiction with the definition of ξ. This proves that $\xi = 1$ and the uniqueness follows.

(Existence) Let Ξ be the set of all $s \in [0, 1]$ such that F is invertible along $\sigma|_{[0,s]}$ with inverse

$$\theta_s : [0, s] \to M_0 \text{ such that } \theta_s(0) = u^*, \ F(u^*) = v^* = \sigma(0).$$

We are going to show that Ξ is open and closed in $[0, 1]$, so that, Ξ containing at least the point 0, we will have that $\Xi = [0, 1]$. To prove that Ξ is closed, we set $\xi = \sup \Xi$, and note that, as before, $\xi > 0$. By uniqueness, the paths θ_s coincide in the intersections of their intervals of definition; let θ denote the function that they define on $[0, \xi)$.

Next, let $s_n \uparrow \xi$ be such that $\sigma(s_n) \to v$. Since $\theta(s_n) = F^{-1}(\sigma(s_n))$ and F is proper we have that (without relabelling) $\theta(s_n) \to u$, with $F(u) = v$. Let U and V be two neighbourhoods of u and v, respectively, such that $F|_U \in \mathrm{Hom}(U, V)$. If $m \in \mathbf{N}$ is such that

$$\theta(s_m) \in U, \text{ and } \sigma([s_m, \xi]) \subset V$$

then F can be inverted along $\sigma|_{[s_m, \xi]}$, giving a path θ_1 such that $\theta_1(s_m) = \theta(s_m)$. According to Remark 1.10 (ii), we can say that F is invertible along $\sigma|_{[0,\xi]}$ and this shows that Ξ is closed. The same construction used above can be employed to prove that Ξ is open. Indeed, if $\xi < 1$, the path θ_1 just introduced could be defined in an interval $[s_m, \xi + \alpha]$, for some $\alpha > 0$ sufficiently small, in contradiction with the assumption that $\xi = \sup \Xi$. This completes the proof of Lemma 1.11.

Let $Q = [0, 1] \times [0, 1]$ and let θ and σ be continuous maps (which we will call "2-paths"):

$$\theta : Q \to M, \ \sigma : Q \to N;$$

as before we say that θ inverts F along σ if

$$F \circ \theta = \sigma.$$

The following lemma is analogous to Lemma 1.11.

Lemma 1.12 *Let* $u^* \in M_0$ *and* $v^* = F(u^*) \in N_0$. *Then, given any 2-path* $\sigma : Q \to N_0$ *such that* $\sigma(0,0) = v^*$, *there exists a unique 2-path* $\theta : Q \to M_0$ *that inverts* F *along* σ *such that* $\theta(0,0) = u^*$.

Proof (Uniqueness). Let $\theta_1, \theta_2 : Q \to M_0$ be 2-paths that invert F

along σ with $\theta_1(0,0) = \theta_2(0,0) = u^*$, and let (s,t) be a generic point of Q. Define $\phi_{1,2} : Q \to M_0$ and $\psi : Q \to N_0$ by setting

$$\phi_1(\lambda) = \theta_1(\lambda s, \lambda t),$$
$$\phi_2(\lambda) = \theta_2(\lambda s, \lambda t),$$
$$\psi(\lambda) = \sigma(\lambda s, \lambda t).$$

Plainly, ϕ_1 and ϕ_2 are paths which invert F along ψ; since $\phi_1(0) = \phi_2(0) = u^*$ and $\psi(0) = v^*$, Lemma 1.11 implies that $\phi_1 = \phi_2$. In particular, setting $\lambda = 1$ we have

$$\theta_1(s,t) = \theta_2(s,t).$$

Let us note that, evidently, the result can be formulated substituting for Q any rectangle R: if θ_1 and θ_2 are 2-paths $R \to M_0$ that invert F along $\sigma|_R$ and if at some $(s^*, t^*) \in R$ one has $\theta_1(s^*, t^*) = \theta_2(s^*, t^*)$, then $\theta_1 = \theta_2$ in R.

(Existence) Consider a rectangle

$$R_s = [0,s] \times [0,1] \subset Q,$$

and let Ξ be the set of all $s \in (0,1]$ such that there exists $\theta_s : R_s \to M_0$ that inverts F along $\sigma|_{R_s}$ with $\theta_s(0,0) = u^*$. Clearly $0 \in \Xi$; indeed, F is invertible along the path $t \to \sigma(0,t)$, in view of Lemma 1.11. Let $\xi = \sup \Xi$; we will show again that $\xi \in \Xi$ and that $\xi = 1$. By uniqueness, all the 2-paths θ_s coincide in the intersection of their domains of definition; thus we can define a 2-path

$$\theta : [0,\xi) \times [0,1] \to M_0$$

which coincides with θ_s in R_s. Let us fix any $t \in [0,1]$; since F is invertible along the path $s \to \sigma(s,t)$ through a path $s \to \phi(s)$, $s \in [0,1]$, such that $\phi(0) = \theta(0,t)$, by uniqueness we have

$$\phi(z) = \theta(z,t) \text{ for all } 0 \leq z < \xi.$$

If we set $\phi(\xi) = u$ and $\sigma(\xi,t) = v$, there exist neighbourhoods U, V of u and v, respectively, such that F induces a homeomorphism between U and V. Then we can find a rectangle $\tilde{R}$ (Fig. 3.1) centred on (ξ,t) and a $\tilde{\theta} : \tilde{R} \cap Q \to M_0$ such that $\tilde{\theta}$ inverts F along $\sigma \tilde{R} \cap Q$ with $\tilde{\theta}(\xi,t) = u$. Since $\tilde{\theta}(z,t) = \theta(z,t)$ for $0 < z < \xi$ (because $\tilde{\theta}(z,t) = \phi(z)$, $\theta(z,t) = \phi(z)$ for $z < \xi$) we infer that θ can be extended to all $\tilde{R} \cap Q$. In this way θ can be extended continuously to all R_ξ in such a way that the relationship $F \circ \theta = \sigma$ holds true therein. Moreover, one must have $\xi = 1$; otherwise, if $\xi < 1$, we could cover the segment $\{(\xi,t) : t \in [0,1]\}$ with a finite family of rectangles like $\tilde{R}$, and θ could be extended to a rectangle $R_{\xi+\alpha}$ with $\alpha > 0$ small enough. This completes the proof of the lemma.

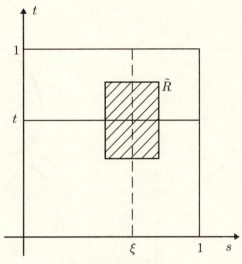

Figure 3.1

Proof of Theorem 1.7.   First from Theorem 1.6 we infer that $[v]$ is (constant and) ≥ 1 for all $v \in N_0$. Hence F is onto N_0.

It remains to show that $[v] = 1$ for all $v \in N_0$. Arguing by contradiction, let $u_0, u_1 \in M_0$ and $v \in N_0$ be such that $F(u_0) = F(u_1) = v$ (Fig.3.2).   Since M_0 is arcwise connected, we can find a path $\theta \in C([0,1], M_0)$ with $\theta(0) = u_0, \theta(1) = u_1$.

The image of θ through F, $\sigma = F \circ \theta$, is a closed curve in the simply connected space N_0. Hence there is a continuous homotopy $h \in C(Q, N_0)$ satisfying (1.4). Without loss of generality, we can assume that $h(s, 1)$ coincide, just with v and that

$$h(0,t) = h(1,t) = v \text{ for all } t \in [0,1]. \tag{1.5}$$

From Lemma 1.12 we infer their exists a a unique 2-path $\Theta \in C(Q, M_0)$ that inverts F along h; that is, such that

$$\Theta(0,0) = u_0,$$

$$F(\Theta(s,t)) = h(s,t), \text{ for all } (s,t) \in Q.$$

In particular, from $F(\Theta(s,0)) = h(s,0) = \sigma(s)$, we deduce that $\Theta(s,0) = \sigma(s)$, hence

$$\Theta(1,0) = \theta(1) = u_1. \tag{1.6}$$

On the other hand, from (1.5) it follows that

$$F(\Theta(0,t)) = h(0,t) = v,$$

$$F(\Theta(s,1)) = h(s,1) = v,$$

$$F(\Theta(1,t)) = h(1,t) = v.$$

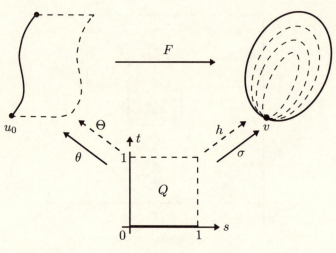

Figure 3.2

Therefore, letting

$$\Gamma = (\{0\} \times [0,1]) \cup ([0,1] \times \{1\}) \cup (\{1\} \times [0,1]),$$

one has $\Theta|_\Gamma =$ constant. In particular

$$\Theta(1,0) = \Theta(0,0) = u_0,$$

in contradiction with (1.6). This shows that $[v] = 1$, and completes the proof of the theorem.

An application

Postponing other applications to the next chapter, we will discuss briefly here a result concerning a class of asymptotically linear Dirichlet boundary value problems like

$$-\Delta u(x) = p(u(x)) + h(x), \quad x \in \Omega,$$
$$u(x) = 0, \quad x \in \partial\Omega,$$

where Ω is a bounded domain in $\mathbb{R}^n$.

We will use the notation introduced in subsection 0.6. In particular, λ_1 will denote the first eigenvalue of

$$-\Delta u = \lambda u \text{ in } \Omega, \; u = 0 \text{ on } \partial\Omega,$$

with associated eigenfunction ϕ_1, with $\phi_1(x) > 0$ in Ω.

Theorem 1.13   *Let* $p \in C^1(\mathbb{R})$ *satisfy*

(1) $p(s) \geq 0$ *for all* s;
(2)   *there exist* $\gamma < \lambda_1$ *and* $b \geq 0$ *such that* $p(s) \leq \gamma s + b$ *for all* $s \geq 0$;
(3) $p'(s) < \lambda_1$.

Then (1.7) *has a unique solution* $u \in C^{2,\alpha}(\overline{\Omega})$ *for any* $h \in C^{0,\alpha}(\overline{\Omega})$.

In order to apply the Global Inversion Theorem 1.8 let

$$X = \{u \in C^{2,\alpha}(\overline{\Omega}) : u(x) = 0 \text{ on } \partial\Omega\},$$
$$Y = C^{0,\alpha}(\overline{\Omega}),$$
$$F : X \to Y, \ F(u) = \Delta(u) + p(u).$$

Plainly, $F \in C^1(X, Y)$ and, given $h \in Y$, any $u \in X$ such that $F(u) = h$ is a solution of (1.7). Let us show

Lemma 1.14   *F is locally invertible on all of* X.

Proof. Since $F'(u) : v \to \Delta v + p'(u)v$, then, according to Theorem 0.7, $F'(u) \in \text{Inv}(X, Y)$ whenever the Dirichlet problem

$$-\Delta v = p'(u)v \text{ in } \Omega, \ v = 0 \text{ on } \partial\Omega$$

has only the trivial solution. This is actually the case because assumption (3) and the comparison property of the eigenvalues (see Theorem 0.6(ii)) imply that $\lambda_1(p'(u)) > 1$.

Lemma 1.15   *F is proper.*

Proof. Let $h_n \in Y$ and $u_n \in X$ be sequences such that

$$\|h_n\|_Y \leq H \text{ and } F(u_n) = -h_n.$$

We claim

(a)   *there exists* $c_1 > 0$ *such that* $u_n(x) \geq -c_1$ *for all* $x \in \Omega$.

To see this, let Ω^* be a bounded domain of $\mathbb{R}^n$ such that $\overline{\Omega} \subset \Omega^*$. Let λ^* denote the first eigenvalue of $-\Delta$ on Ω^* with zero Dirichlet boundary conditions and let ϕ^* be such that $\phi^*(x) > 0$ on Ω^* and

$$-\Delta\phi^* = \lambda^*\phi^* \text{ in } \Omega^*, \ \phi^* = 0 \text{ on } \partial\Omega^*.$$

Since $\overline{\Omega} \subset \Omega^*$ and $\phi^*(x) > 0$ on Ω^* then $\mu^* = \min_{x \in \overline{\Omega}} \phi^*(x) > 0$. Let us set $c^* = H \cdot (\lambda^*\mu^*)^{-1}$. Using assumption (a) and the definition of c^* one readily finds

$$-\Delta(u_n + c^*\phi^*) = p(u_n) + h_n + c^*\lambda^*\phi^*$$
$$\geq h_n + c^*\lambda^*\phi^* \geq h_n + H \geq 0.$$

Since, in addition, $u_n(x) + c^*\phi^*(x) > 0$ for all $x \in \partial\Omega$, the Maximum

Principle 0.8 implies that $u_n \geq -c^*\phi^*$ in Ω, and this proves the claim (a).

Next we show

(b)   *there exists $c_2 > 0$ such that $u_n(x) \leq c_2$ for all $x \in \Omega$.*

Indeed, let $K = H + b$ and let w be a solution of

$$-\Delta w = \gamma w + K \text{ in } \Omega, \; w = 0 \text{ on } \partial\Omega.$$

Note that such a solution exists because $\gamma < \lambda_1$ and $w > 0$ in Ω, by the Maximum Principle. Setting $z_n = w - u_n$, one finds

$$-\Delta z_n = \gamma w + K - p(u_n) - h_n \text{ in } \Omega.$$

Let $\Omega_n^+ := \{x \in \Omega : u_n(x) > 0\}$. Using assumption (2) one readily infers

$$-\Delta z_n \geq \gamma w + K - \gamma u_n - b - h_n \geq \gamma z_n \text{ in } \Omega_n^+.$$

Moreover, $z_n(x) = w(x) > 0$ for all $x \in \partial\Omega_n^+$.

Now, Theorem 0.6(v) implies that $\lambda_1(\Omega_n^+) \geq \lambda_1$ and therefore, since $\gamma < \lambda_1$, the Maximum Principle 0.8 applies to z_n yielding $z_n(x) \geq 0$ in Ω_n^+. As $z_n(x) = z(x) - u_n(x) > 0$ in $\Omega - \Omega_n^+$, it follows that $z_n \geq 0$ in Ω and this proves (b).

From the preceding steps (a) and (b) it follows that $\|u_n\|_\infty \leq c_3$. Since u_n satisfies

$$-\Delta u_n = p(u_n) + h_n,$$

a repeated application of Theorem 0.5 (ii) and (iii) implies $\|u_n\|_X \leq c_4$. Then, without relabelling, u_n converges in $C^2(\overline{\Omega}), p(u_n) + h_n$ converges in Y and finally, using again equation (1.8), u_n converges in X. This completes the proof of the lemma.

Lemmas 1.14 and 1.15 allow us to apply Theorem 1.8 to F and this suffices to prove Theorem 1.13.

2 Global inversion with singularities

The main purpose of this section is to study the global invertibility of mappings when the singular set Σ is such that Theorem 1.7 does not apply. Our goal will be a global, rather precise, geometric description of Σ, of $F(\Sigma)$ and of the image of F. For this, it will be convenient to deal with smooth (C^2) maps $F : X \to Y$ where X and Y are Banach spaces, and substitute for Σ the set

$$\Sigma' := \{u \in X : F'(u) \notin \text{Inv}(X, Y)\}.$$

Clearly $\Sigma' \supset \Sigma$.

Let $F \in C^2(X, Y)$, $u \in \Sigma'$ and suppose

(a) Ker$(F'(u))$ is one-dimensional: let $\phi \in X - \{0\}$ be such that Ker$(F'(u)) = \mathbb{R}\phi$; $R(F'(u))$ is closed and has codimension one;

(b) There exists $\tilde{\phi} \in X$ such that $F''(u)[\tilde{\phi}, \phi] \notin R(F'(u))$.

Recall that a subset M of X is said to be *a C^1-manifold of codimension 1 in X* if for all $u^* \in M$ there exist $\delta > 0$ and a functional $\Gamma : B_\delta(u^*) \to \mathbb{R}$ of class C^1 such that

$$M \cap B_\delta(u^*) = \{u \in B_\delta(u^*) : \Gamma(u) = 0\}, \tag{2.1}$$

$$\Gamma'(u^*) \neq 0. \tag{2.2}$$

We anticipate that *if M is a closed, connected, C^1-manifold of codimension 1 in a Banach space X, then $X \backslash M$ has at most two components.* The proof of this fact is postponed to the appendix.

First of all we prove a lemma.

Lemma 2.1 *Suppose for all $u \in \Sigma'$ conditions (a)–(b) hold. Then Σ is a C^1 manifold of codimension 1 in X.*

Proof. Fixing an arbitrary $u^* \in \Sigma'$, we shall describe Σ' in a neighbourhood of u^*. Let

$$V = \text{Ker}(F'(u^*)), R = R(F'(u^*)).$$

From (a) it follows that there exists $\phi \in X - \{0\}$ and $\psi \in Y^* - \{0\}$ (depending on u^*) such that $V = \mathbb{R}\phi$ and $R = \text{Ker}(\psi)$, and there exist linear subspaces W (resp.Z) in X (resp.Y) such that

$$X = V \oplus W, \ Y = Z \oplus R.$$

For any $u \in X$ there are unique $t \in \mathbb{R}$ and $w \in W$ such that

$$u = t\phi + w.$$

Moreover we let Q and $P = I - Q$ denote the projections onto R and Z, respectively. Hence, if $z \in Z$ is such that $\langle \psi, z \rangle = 1$, then $Pv = \langle \psi, v \rangle z$, for all $v \in Y$.

For u near u^* we want to see whether $F'(u) \in \text{Inv}(X, Y)$ or not. So we consider the equation $F'(u)(t\phi + w) = v$. Applying P and Q we find

$$\left. \begin{array}{l} PF'(u)(t\phi + w) = Pv, \\ QF'(u)(t\phi + w) = Qv. \end{array} \right\} \tag{2.3}$$

For $u = u^*$ one has $F'(u^*)\phi = 0$ and the second of (2.3) becomes

$$QF'(u^*)w = Qv.$$

Now $QF'(u^*)$ is invertible, as a linear bounded map from W to R. Since $\text{Inv}(W, R)$ is open (see Proposition 2.1.1), we can find $\delta > 0$ such that

$$QF'(u) \in \text{Inv}(W, R), \text{ for all } u \in B_\delta(u^*).$$

Setting $T = [QF'(u)]^{-1}$, one has $w = T[Qv - tQF'(u)\phi]$ and the first of (2.3) yields

$$tPF'(u)\phi + PF'(u)T[Qv - tQF'(u)\phi] = Pv.$$

Then (2.3) is equivalent to the system

$$t\langle\psi, F'(u)\phi\rangle - t\langle\psi, F'(u)TQF'(u)\phi\rangle = \langle\psi, v\rangle - \langle\psi, F'(u)TQv\rangle, \quad (2.4)$$

$$w = T[Qv - tQF'(u)\phi]. \qquad (2.5)$$

Since (2.4) is uniquely solvable whenever

$$\langle\psi, F'(u)\phi\rangle - \langle\psi, F'(u)TQF'(u)\phi\rangle \neq 0,$$

it follows that u is singular if and only if

$$\langle\psi, F'(u)\phi\rangle - \langle\psi, F'(u)TQF'(u)\phi\rangle = 0.$$

Hence letting

$$\Gamma(u) = \langle\psi, F'(u)\phi\rangle - \langle\psi, F'(u)TQF'(u)\phi\rangle,$$

we get

$$u \in \Sigma' \cap B_\delta(u^*) \Leftrightarrow \Gamma(u) = 0, \ u \in B_\delta(u^*).$$

Since Γ is obivously C^1, it remains to show that (2.2) holds. In fact, with easy calculations one finds

$$\Gamma'(u^*)u = \langle\psi, F''(u^*)[u, \phi]\rangle. \qquad (2.6)$$

Therefore (b) yields

$$\Gamma'(u^*)\tilde{\phi} = \langle\psi, F''(u^*)[\tilde{\phi}, \phi]\rangle \neq 0,$$

proving the lemma.

To describe $F(\Sigma')$ we shall strengthen (b), giving the following definition.

Definition 2.2 We say that $u \in \Sigma'$ is an *ordinary* singular point if (a) holds and

(c) $F''(u)[\phi, \phi] \notin R(F'(u)),$

where, according to (a), $\phi \neq 0$ is such that $\mathrm{Ker}(F'(u)) = \mathbb{R}\phi$.

Lemma 2.3 *Let u^* be an ordinary singular point. Then there exist $\varepsilon > 0$ and a map $\Psi \in C^1(B_\varepsilon(u^*), Y)$ such that*

(i) $\Psi'(u^*) \in Inv(X, Y),$

(ii) $\Psi(u) = F(u)$ *for all* $u \in \Sigma' \cap B_\varepsilon(u^*).$

Proof. We will keep the notation introduced in Lemma 2.1. From that lemma it follows that $\Sigma' \cap B_\delta(u^*) = \Gamma^{-1}(0)$. Let $\Psi : B_\delta(u^*) \to Y$ be defined by

$$\Psi(u) = F(u) + \Gamma(u)z.$$

The map Ψ is C^1 and

$$\Psi(u) = F(u), \text{ for all } u \in \Sigma' \cap B_\delta(u^*).$$

Moreover, there results

$$\Psi'(u^*)u = F'(u^*)u + \Gamma'(u^*)(u)z.$$

Setting $u = t\phi + w$, and using (2.6) we find

$$\Psi'(u^*)u = F'(u^*)w + t\Gamma'(u^*)(\phi)z + \Gamma'(u^*)(w)z$$
$$= F'(u^*)w + t\langle \psi, F''(u^*)[\phi, \phi]\rangle z + \langle \psi, F''(u^*)[w, \phi]\rangle z.$$

It is readily verified that $\Psi'(u^*)u = v$ has a unique solution whenever $\langle \psi, F''(u^*)[\phi, \phi]\rangle \neq 0$. Hence, if (c) holds, $\Psi'(u^*) \in \text{Inv}(X, Y)$ and (i) follows.

Corollary 2.4 *If every $u \in \Sigma'$ is an ordinary singular point, then $F(\Sigma')$ is a C^1-manifold of codimension 1 in Y.*

Proof. From the lemma we can find an $\varepsilon > 0$ and a neighbourhood N of $F(u)$ such that Ψ induces a diffeomorphism between $B_\varepsilon(u)$ and N. Plainly, the functional

$$\gamma := \Gamma \circ \Psi^{-1} : N \to \mathbb{R}$$

is C^1 and has non-zero derivative. Moreover, Lemma 2.3 (ii) implies immediately that $F(\Sigma') \cap N = \gamma^1(0)$, proving the corollary.

Assumption (c) allows us to evaluate the "local" number of the solutions of $F(u) = v$. More precisely one has the following

Lemma 2.5 *Let u^* be an ordinary singular point with $\text{Ker}(F'(u^*)) = \mathbb{R}\phi$, and, say,*

$$\langle \psi, F''(u^*)[\phi, \phi]\rangle > 0,$$

and set $v^* = F(u^*)$. Then there are $\varepsilon, \sigma > 0$ such that the equation $F(u) = v^* + sz, u \in B_\varepsilon(u^*)$, has

$$\left. \begin{array}{l} \text{two solutions for all } 0 < s < \sigma, \\ \text{no solutions for all } -\sigma < s < 0. \end{array} \right\}$$

Proof. (see Figure 3.3) For simplicity of notation we take $u^* = 0$ and $F(u^*) = 0$ and study the equation $F(u) = sz$ in a neighbourhood of 0.

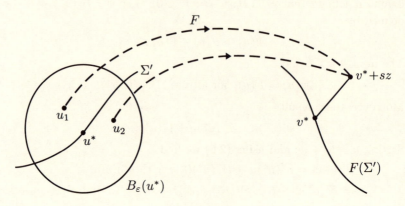

Figure 3.3

Set $A = F'(0)$ and $F(u) = Au + \omega(u)$. Substituing $u = t\phi + w$ in $F(u) = sz$ we find

$$Aw + \omega(t\phi + w) = sz.$$

Applying the projections P and Q we find the equivalent system

$$\left.\begin{array}{r} Aw + Q\omega(t\phi + w) = 0, \\ P\omega(t\phi + w) = sz. \end{array}\right\} \tag{2.7}$$

This procedure (already used, in a linear framework, in Lemma 2.1) is usually referred as "Liapunov-Schmidt reduction" and will be discussed in greater generality in Section 5.3.

Since $\omega(0) = 0$, $\omega'(0) = 0$ and $A \in \mathrm{Inv}(W, R)$, we can apply the Implicit Function Theorem to $Aw + Q\omega(t\phi + w) = 0$, yielding solutions $w = w(t)$, with w of class C^2 such that $w(0) = 0$ and $w'(0) = 0$.

Inserting in the second of (2.7) we are led to solve

$$\chi(t) := \langle \psi, \omega(t\phi + w(t)) \rangle = s.$$

The map χ is C^2 (because ω and w are) and there results (we put $u_t = t\phi + w(t)$)

$$\chi'(t) = \langle \psi, \omega'(u_t)[\phi + w'(t)] \rangle$$

$$\chi''(t) = \langle \psi, \omega''(u_t)[\phi + w'(t), \phi + w'(t)] \rangle + \langle \psi, \omega'(u_t)[w''(t)] \rangle.$$

Since $w(0) = w'(0) = 0$, $\omega'(0) = 0$ and $\omega''(0) = F''(0)$, we deduce

$$\chi'(0) = 0, \chi''(0) = \langle \psi, F''(0)[\phi, \phi] \rangle > 0,$$

and the lemma follows.

We are now in position to state the main result of this section.

Theorem 2.6 *Suppose that*

(1) $F \in C^2(X, Y)$ *and is proper,*
(2)   *every $u \in \Sigma'$ is an ordinary singular point,*
(3)   *for all $v \in F(\Sigma')$ the equation $F(u) = v$ has a unique solution,*
(4) Σ' *is connected.*
Then there exist two open connected subsets Y_0 and Y_2 such that
(i) $Y = Y_0 \cup Y_2 \cup F(\Sigma')$,
(ii) *the number $[v]$ of solutions of the equation $F(u) = v$ is*

$$[v] = \begin{cases} 0 & \text{if } v \in Y_0, \\ 1 & \text{if } v \in F(\Sigma'), \\ 2 & \text{if } v \in Y_2. \end{cases}$$

In order to carry out the proof of Theorem 2.6 we need a further lemma.

Lemma 2.7 *Let $u \in \Sigma'$. Then for every neigbourhood U of u there exists a neighbourhood V of $F(u)$ such that $F^{-1}(V) \subset U$.*

Proof. Otherwise we could find a neigbourhood U^* of u and a sequence $u_n \notin U^*$ such that $F(u_n) \to F(u)$. Since F is proper, one has (without relabelling) that $u_n \to u^* \notin U^*$, with $F(u^*) = F(u)$. This is in contradiction with assumption (3).

Proof of Theorem 2.6 From Corollary 2.4 $F(\Sigma')$ is a C^1-manifold of codmension 1 in Y. According to assumption (4) Σ' and (hence) $F(\Sigma')$ are connected. The result mentioned before (see appendix), implies that $Y \backslash F(\Sigma')$ consists at most of two connected components. Let $u^* \in \Sigma'$ be fixed and let $B_\varepsilon(u^*)$ be the neighbourhood found in Lemma 2.5. Applying Lemma 2.7 with $U = B_\varepsilon(u^*)$ we infer that for all $v \in V$ the number $[v]$ equals the "local" number of solutions of $F(u) = v$, with $u \in U$, a number that has been evaluated in Lemma 2.5. It follows that $[v]$ can be either zero or 2. Moreover, $[v]$ is constant on each component of $Y \backslash f(\Sigma')$, which therefore consists exactly of two connected components Y_0 and Y_2, say, with the properties listed in (ii).

An application of Theorem 2.6 will be discussed in Section 2 of the next chapter.

Appendix

Here we prove the following.

Proposition *Let M be a closed connected C^1-manifold of codimension 1 in the Banach space X. Then $X \backslash M$ has at most two connected components.*

Proof. Supposing the contrary, let A_1, A_2, A_3 be nor empty, open, disjoint subsets of X such that $X \backslash M = A_1 \cup A_2 \cup A_3$. Since $X \backslash M$ is open, then each A_i is open with respect to X, too. Let Γ_i denote the boundary of A_i. For any $i = 1, 2, 3$ one has that $\Gamma_i \neq \emptyset$ (otherwise A_i would also be closed), is closed and Γ_i is contained in M. Since M is a C^1-manifold of codimension 1, for any $u \in M$ there is an $\varepsilon > 0$ such that $B_\varepsilon(u) \cap (X \backslash M)$ consists exactly of two components, say U_1 and U_2. As a consequence, only two of the A_i can have non-empty intersection with $B_\varepsilon(u)$ and thus $B_\varepsilon(u) \cap M$ can be contained in two of the Γ_i, at most. Let, for example, U_1 be contained in A_1 and U_2 in A_2. It follows immediately that

$$\Gamma_3 \cap [B_\varepsilon(u) \cap M] = \emptyset. \qquad (A1)$$

Now, let v be any point of Γ_3. As before, there is a $\delta > 0$ such that $B_\delta(v) \cap [X \backslash M]$ consists of two components. Since $v \in \Gamma_3$, one of those components has to be contained in A_3. Thus all $w \in B_\delta(v) \cap M$ belong to Γ_3 and hence Γ_3 is (closed and) open. Since M is connected, $M = \Gamma_3$, in contradiction with (A1).

4

Semilinear Dirichlet problems

In this chapter we will apply the global inversion theorems discussed in the preceding chapter to the study of some classes of semilinear elliptic Dirichlet boundary-value problems with asymptotically linear nonlinearity. In many cases the results concerning the existence of solutions could be obtained by means of other tools (for example, the topological degree). In the spirit of what we discussed in Chapter 3 we prefer to use the inversion theorems. According to their specific nature, these abstract theorems lead to the finding of precise results under suitable restrictions. However we shall see that the Global Inversion Theorem 3.1.8, used in conjunction with the Lyapunov—Schmidt procedure (see Section 3.2) lets us handle a broad class of "Problems at Resonace" which are discussed in Section 1. Let us note that the functional setting is that of the Hölder spaces and the results are founds with rather simple arguments, which are geometrically expressive.

In Section 2 we deal with the so-called "jumping nonlinearities" which are handled by means of Theorem 3.2.6. According to that theorem we can establish the *precise number* of solutions for this class of boundary-value problems. In addition, we show how the existence results can be completed by using other tools, such as the method of "sub-" and "super-solutions".

We are aware that our discussion is far from complete and refer to Remarks 1.9 and 1.10 as well as to the cited papers and bibliography therein for many other results on asymptotically linear Dirichlet problems.

Notation

Throughout this chapter, Ω denotes a bounded domain in $\mathbb{R}^n$ with smooth boundary $\partial\Omega$.

1 Problems at resonance

Consider the semilinear Dirichlet problem

$$\left.\begin{array}{r}\Delta u + p(u) = h(x) \text{ in } \Omega,\\ u = 0 \text{ on } \partial\Omega,\end{array}\right\} \tag{D}$$

where p is asymptotically linear, and $h \in C^{0,\alpha}(\overline{\Omega})$.

More precisely, in the present section we consider nonlinearities p of the form

$$p(s) = as + b(s)$$

with $a \in \mathbb{R}$, and $b \in C(\mathbb{R})$ bounded. See Remark 1.13 below for a slightly more general class of problems.

We shall distinguish between the cases $a = \lambda_k$ or $a \neq \lambda_k$, where λ_k denotes the k-th eigenvalue of $-\Delta$ with zero Dirichlet boundary conditions (see Subsection 0.6).

Case $a \neq \lambda_k$ for all k

Let us by recalling that in the present case (D) is always solvable.

Theorem 1.1 *If $a \neq \lambda_k$ for all k, and b is Lipschitz-continuous and bounded, then for all $h \in C^{0,\alpha}(\overline{\Omega})$ (D) has a solution $u \in C^{2,\alpha}(\overline{\Omega})$.*

For a proof we refer to [**Maw**]. Let us recall that a main tool is an *a priori* estimate like that stated in Lemma 1.2 below.

We want to prove below an *existence and uniqueness result* by means of the Global Inversion Theorem 3.1.8.

We shall work in Hölder spaces (similar arguments could carried over taking Sobolev spaces). Set

$$X = \{u \in C^{2,\alpha}(\overline{\Omega}) : u = 0 \text{ on } \partial\Omega\}, Y = C^{0,\alpha}(\overline{\Omega}),$$

and consider the map F defined on X by

$$F(u) = \Delta u + au + b(u).$$

Plainly, F maps X into Y and is continuous. Let us start with an *a priori* bound for the solutions of (D).

Lemma 1.2 *Suppose $a \neq \lambda_k$ for all k, and b is bounded, and let $h_n \in$*

$Y, u_n \in X$ be such that $F(u_n) = h_n$. Then $\|u_n\|_Y = \|u_n\|_{C^{0,\alpha}}$ is bounded provided $\|h_n\|_Y$ is.

Proof. If not, let $z_n(x) = u_n(x)/\|u_n\|_Y$. Dividing the equation $F(u_n) = h_n$ by $\|u_n\|_Y$ we find

$$\Delta z_n + a\, z_n + \frac{b(u_n(x))}{\|u_n\|_Y} = \frac{h_n}{\|u_n\|_Y}. \tag{1.1}$$

Set $U_n := -b(u_n(x))/\|u_n\|_Y + h_n/\|u_n\|_Y$; since U_n is bounded in Ω, Theorem 0.5 (ii) implies that $\|z_n\|_{C^{1,\alpha}} \leq$ const. and, up to a sub-sequence, $z_n \to z^*$ in $C^1(\overline{\Omega})$. Since $\|z_n\|_Y = 1$, then $\|z^*\|_Y = 1$; in particular z^* is not identically zero.

Multiplying (1.1) by $w \in C_0^\infty(\Omega)$ and integrating we find

$$-\int_\Omega \nabla w \nabla z_n + a \int_\Omega w z_n = \int_\Omega w U_n. \tag{1.2}$$

In the right-hand side one has that $U_n \to 0$ (uniformly) because $\|h_n\|_Y \leq c_1$, b is bounded and $\|u_n\|_Y \to \infty$. Then, passing to the limit as $n \to \infty$ in (1.2) we get

$$-\int_\Omega \nabla w \nabla z^* + a \int_\Omega w z^* = 0, \text{ for all } w \in C_0^\infty(\Omega).$$

This means that weakly, and by Theorem 0.5 (iii), strongly, $\Delta z^* + a z^* = 0$. This is a contradiction, because $a \neq \lambda_k$.

From Lemma 1.2 we deduce the following.

Corollary 1.3 *The map $F : X \to Y$ is proper.*

Proof. Let $h_n \to h$ in Y and $u_n \in X$ be such that $F(u_n) = h_n$. From the preceding lemma one has that $\|u_n\|_Y \leq c_1$. Setting $\theta_n = -a u_n - b(u_n) + h_n$ we find that $\|\theta_n\|_Y \leq c_2$. Since $-\Delta u_n = \theta_n$, Theorem 0.5 (iii) yields $\|u_n\|_X \leq c_3$. Therefore $u_n \to u^*$ in $C^2(\overline{\Omega})$ (up to a sub-sequence). As a consequence, θ_n converges in $C^{0,\alpha}$ and $u_n \to u^*$ in X, proving the claim.

We are now in position to state the following result.

Theorem 1.4 *Suppose that*

(b1) *$b \in C^1(\mathbb{R})$ and there exists $M > 0$ such that $|b(s)| \leq M$ for all $s \in \mathbb{R}$,*

(b2) *either $a + b'(s) < \lambda_1$ for all $s \in \mathbb{R}$, or $\lambda_k < a + b'(s) < \lambda_{k+1}$ for all $s \in \mathbb{R}$.*

Then for all $h \in C^{0,\alpha}(\overline{\Omega})$ *problem* (D) *has a unique, classical, solution* $u \in C^{2,\alpha}(\overline{\Omega})$.

Proof. Note that now F is of class C^1. Moreover one has

$$F'(u) : v \to \Delta v + av + b'(u)v.$$

In order to apply the Global Inversion Theorem 3.1.8 to F, it remains to prove the following lemma which is similar to Lemma 1.14 of Chapter 3.

Lemma 1.5   *F is locally invertible on X.*

Proof. Let $u \in X$ and set $m(x) = au(x) + b'(u(x))$. From (b2) it follows that

$$\lambda_k < m < \lambda_{k+1} \text{ (or } m < \lambda_1)$$

and the comparison property of eigenvalues (Theorem 0.6 (ii)) yields

$$\lambda_k(m) < 1 < \lambda_{k+1}(m) \text{ (resp. } \lambda_1(m) > 1)$$

and this means that the linear b.v.p.

$$\left.\begin{array}{r} -\Delta v = mv \text{ in } \Omega, \\ v = 0 \text{ on } \partial\Omega, \end{array}\right\}$$

has the trivial solution $v \equiv 0$ only. According to Theorem 0.7 this suffices to show that $F'(u)$ is invertible.

Proof of Theorem 1.4 completed . Lemma 1.5 and Corollary 1.3 allow us to apply the Global Inversion Theorem 3.1.8.

Case $a = \lambda_k$ (Problems at resonance)

If $a = \lambda_k$ then problem (D) becomes

$$\left.\begin{array}{r} \Delta u + \lambda_k u + b(u) = h \text{ in } \Omega, \\ u = 0 \text{ on } \partial\Omega, \end{array}\right\} \qquad \text{(PR)}$$

which is usually called *problem at resonance*. It is clear that, in contrast with the results stated in Theorems 1.1–1.3, (PR) may now have no solution at all: this is indeed the case if $b \equiv 0$ and $\int_\Omega h\phi_k \neq 0$. It is easy to extend this non-existence result to the semilinear case. Let λ_k be simple and (for a fixed ϕ_k) set

$$\Omega^+ = \{x \in \Omega : \phi_k(x) > 0\}, \qquad \Omega^- = \{x \in \Omega : \phi_k(x) < 0\}, \qquad (1.3)$$

$$m^- = \inf\{b(s) : s \in \mathbb{R}\}, \qquad m^+ = \sup\{b(s) : s \in \mathbb{R}\},$$

and

$$A = m^- \int_{\Omega^+} \phi_k + m^+ \int_{\Omega^-} \phi_k, \quad B = m^+ \int_{\Omega^+} \phi_k + m^- \int_{\Omega^-} \phi_k.$$

Proposition 1.6 *Let λ_k be simple and let $b(s)$ be bounded. Then a necessary condition for* (PR) *to have a solution is that*

$$A \le \int_\Omega h\phi_k \le B. \tag{1.4}$$

Proof. Let u be a solution of (PR). Multiplying by ϕ_k and integrating one finds

$$\int_\Omega u\Delta\phi_k + \int_\Omega (\lambda_k u + b(u))\phi_k = \int_\Omega h\phi_k.$$

Since $\Delta\phi_k + \lambda_k\phi_k = 0$, we deduce

$$\int_\Omega b(u)\phi_k = \int_\Omega h\phi_k.$$

From this, the result follows immediately.

Let us remark explicitly that (1.4) remains the same if ϕ_k is replaced by $-\phi_k$.

The question whether condition (1.4) is also sufficient for the existence of solutions of (PR) was first investigated by Landesman & Lazer [**LL**]. In the rest of this section we will prove some results concerning this problem, under some restrictive conditions, such as the simplicity of λ_k and (b3) below. These assumptions allow us to use rather elementary arguments and to point out some interesting phenomena. The discussion follows [**AM1**].

Suppose that λ_k is simple, and let W denote the L^2-orthogonal complement of $\mathbb{R}\phi_k$, namely

$$W = \{w \in X : (w|\phi_k) = 0\}.$$

Here and always below we use the notation $(u|v) = \int_\Omega uv$. Every $u \in X$ will be written in a unique way in the form $u = t\phi_k + w$, with $w \in W$ and $t = (u|\phi_k)$. Substituting into (PR) we find readily

$$\Delta w + \lambda_k w + b(t\phi_k + w) = h. \tag{1.5}$$

We will employ the "Lyapunov-Schmidt reduction", as in Lemma 3.2.5. Letting P denote the projection onto W, namely

$$Pu = u - (u|\phi_k)\phi_k(= w),$$

(1.5) turns out to be equivalent to the system

$$\left.\begin{array}{l} \Delta w + \lambda_k w + Pb(t\phi_k + w) = Ph, \\ (b(t\phi_k + w)|\phi_k) = (h|\phi_k). \end{array}\right\} \qquad (1.6)$$

The first equation in (1.6) can be solved as in the non-resonant case. To be precise, let $\Phi : \mathbb{R} \times W \to Y \cap W$ denote the map defined by

$$\Phi(t, w) = \Delta w + \lambda_k w + Pb(t\phi_k + w).$$

Lemma 1.7 *Suppose b satisfies* (b1) *and*

(b3) $\lambda_{k-1} < \lambda_k + b'(s) < \lambda_{k+1} (k > 1)$ *for all $s \in \mathbb{R}$ (or $\lambda_1 + b'(s) < \lambda_2$).*
   Then for all $h \in Y$ there is a unique $w(t, Ph)$ (for brevity we will omit hereafter the dependence on Ph) such that
(i) $\Phi(t, w(t)) = Ph$,
(ii) $t \to w(t)$ *is C^1,*
(iii) $\|w(t)\|_{C^{0,\alpha}} \leq const.$

Proof. We want to show that the Global Inversion Theorem applies to $\Phi(t, .)$. The properness of $w \to \Phi(t, w)$ follows in the same way as for F in Theorem 1.5. In order to show that $\Phi_w(t, w)$ is invertible on W it suffices, according to Theorem 0.7, to prove that the linear homogeneous problem

$$\Delta z + \lambda_k z + Pb'(t\phi_k + w)z = 0, \ z \in W, \qquad (1.7)$$

has the trivial solution $z \equiv 0$ only. Setting $m(x) = b'(t\phi_k(x) + w(x))$, one has that $Pb'(t\phi_k + w) = mz - (mz|\phi_k)\phi_k$ and (1.7) becomes

$$\Delta z + (\lambda_k + m)z - (mz|\phi_k)\phi_k = 0, z \in W. \qquad (1.8)$$

Let us consider the case $k > 1$ (the case $k = 1$ requires some changes in the notation only) and denote by W_1 (resp. W_2) the space spanned by $\{\phi_1, \ldots, \phi_{k-1}\}$ (resp. by $\{\phi_{k+1}, \phi_{k+2}, \ldots\}$), in such a way that $W = W_1 \oplus W_2$, and $z = z_1 + z_2$, with $z_i \in W_i$. Multiplying (1.8) by $z_i (i = 1, 2)$ and integrating we find

$$(z_i|\Delta z) + \lambda_k(z_i|z) + (z_i|mz) = 0, \ i = 1, 2.$$

Setting $z = z_1 + z_2$ since $(z_1|z_2) = (\nabla z_1|\nabla z_2) = 0$, we get

$$-\int_\Omega |\nabla z_i|^2 + \int_\Omega (\lambda_k + m)z_i^2 + \int_\Omega mz_1 z_2 = 0 \ (i = 1, 2)$$

and hence

$$-\int_\Omega |\nabla z_1|^2 + \int_\Omega (\lambda_k + m)z_1^2 = -\int_\Omega |\nabla z_2|^2 + \int_\Omega (\lambda_k + m)z_2^2, \qquad (1.9)$$

If z is not identically zero, (1.9) and (b3) imply

$$\int_\Omega |\nabla z_2|^2 - \int_\Omega |\nabla z_1|^2 = \int_\Omega (\lambda_k + m)(z_2^2 - z_1^2)$$

$$< \lambda_{k+1} \int_\Omega z_2^2 - \lambda_{k-1} \int_\Omega z_1^2. \tag{1.10}$$

Note that $(z_2|\phi_i) = 0$ for all $i = 1, 2, \ldots, k$. Hence from the variational characterization of λ_{k+1} (see Theorem 0.6 (iii)) it follows that

$$\int_\Omega |\nabla z_2|^2 \geq \lambda_{k+1} \int_\Omega z_2^2. \tag{1.11}$$

On the other hand, from $z_1 = \Sigma_{1 \leq i \leq k-1} \zeta_i \phi_i$ one has readily

$$\int_\Omega |\nabla z_1|^2 \leq \lambda_{k-1} \int_\Omega z_1^2. \tag{1.12}$$

Subtracting (1.12) from (1.11) we find a contradiction with respect to (1.10). This proves that $z \equiv 0$. Therefore Φ_w is invertible and (i–ii) follow from Theorem 3.1.8. As for (iii) we can argue as follows. Let us write w_t for $w(t)$. From $\Delta w_t + \lambda_k w_t = Ph - Pb(t\phi_k + w_t)$, and since b is bounded it follows (see Theorem 0.5 (ii)) that $\|w_t\|_{C^{1,\alpha}} \leq$ const. Then $\exists a > 0$ such that $\|Ph - Pb(t\phi_k + w_t)\|_{C^{0,\alpha}} \leq a$. Using Theorem 0.5 (iii) we get the result

Set $\Gamma(t) = \int_\Omega b(t\phi_k + w(t))\phi_k$. According to the previous lemma $\Gamma : \mathbb{R} \to \mathbb{R}$ is continuous. Moreover, the preceding discussion shows that to find a solution of (PR) it suffices to solve the one-dimensional equation

$$\Gamma(t) = \int_\Omega h\phi_k. \tag{1.13}$$

Roughly, we will see that the behaviour of Γ as $|t| \to \infty$ is closely related to that of $b(s)$ as $|s| \to \infty$.

We suppose that

(b4) $b(s) \to b^+(b^-) \in \mathbb{R}$ *as* $s \to +\infty$ (−∞, *respectively*)

and set

$$A' = b^- \int_{\Omega^+} \phi_k + b^+ \int_{\Omega^-} \phi_k, \quad B' = b^+ \int_{\Omega^+} \phi_k + b^- \int_{\Omega^-} \phi_k,$$

where Ω^+ and Ω^- have been defined in (1.3). Without loss of generality, we can take $A' < B'$.

Theorem 1.8 *Suppose λ_k is simple and that b satisfies* (b1), (b3) *and*

(b4). *Then* (PR) *has a solution provided*
$$A' < (h|\phi_k) < B'$$

Proof. Let $t_n \to +\infty$ and set $w_n = w(t_n)$. Using Lemma 1.7 (iii) we infer that w_n converges uniformly to some w^* (up to a sub-sequence). Then

$$t_n \phi_k(x) + w_n(x) \to +\infty(-\infty) \text{ for all } x \in \Omega^+ (\text{resp. } \Omega^-).$$

As b is bounded, an application of the Lebesgue Dominated-Convergence Theorem yields

$$\Gamma(t_n) \to b^+ \int_{\Omega^+} \phi_k + b^- \int_{\Omega^-} \phi_k = B' \quad (t_n \to +\infty).$$

Similarly, one has

$$\Gamma(t_n) \to b^- \int_{\Omega^+} \phi_k + b^+ \int_{\Omega^-} \phi_k = A' \quad (t_n \to -\infty).$$

Since Γ is continuous, it follows that (1.13) has a solution t^* provided (1.4) holds. The corresponding $u^* = t^* \phi_k + w(t^*)$ gives rise to a solution of (PR).

Remarks 1.9

(i) Assumption (b3) and the simplicity of λ_k can be eliminated. Following [AM2] the argument is, roughly, as follows. Let $i, j \in \mathbb{N}$ be such that $\lambda_i < c_1 \le \lambda_k + b'(u) \le c_2 < \lambda_j$ for all $u \in \mathbb{R}$. Set

$$V = \text{span } \{\phi_{i+1}, \dots, \phi_{j-1}\}$$

and let W denote the L^2-orthogonal complement of V with projections Q and P respectively. As before, any $u \in X$ can be written in the (unique) form $u = v + w$, with $v = Qu \in V$ and $w = Pu \in W$ and (PR) can be again replaced by the equivalent system

$$\left. \begin{array}{l} PF(v+w) = Ph, \\ QF(v+w) = Qh. \end{array} \right\}$$

The first equation (on W) can be uniquely solved finding $w = w(v)$, and we are led to the finite-dimensional equation $P(F(v+w(v))) = Ph$. This latter can be studied by means of the Brouwer topological degree.

(ii) For other results on problems at resonance, see for example the book by Fučik [**Fu**] which contains an extensive bibliography. Problems at resonance where the linear part (together with the boundary conditions) gives rise to Fredholm operators with positive index have been investigated in [**Sche**] and [**AAM**].

In the rest of this section we will discuss some of the possible phenomena arising in the study of (PR). The following result deals with a case in which (1.4) is not satisfied.

Theorem 1.10 *Suppose that λ_k is simple and (b1) and (b3) hold. Moreover let us assume that*

(b5) $sb(s) \to \sigma > 0$ *as* $|s| \to \infty$.
  Then (PR) has a solution provided $(h|\phi_k) = 0$.

Proof. We keep the same notation as before. In particular, Lemma 1.7 holds true, and taking h such that $(h|\phi_k) = 0$ we are led to solve the equation $\Gamma(t) = 0$, where Γ is defined in (1.13).
  To use (b5) it is convenient to write $t\Gamma(t)$ in the following form:

$$t\Gamma(t) = \int_\Omega b(t\phi_k + w(t))(t\phi_k + w(t)) - \int_\Omega b(t\phi_k + w(t))w(t).$$

Let t_n be any sequence such that $|t_n| \to \infty$ and set $w_n = w(t_n)$ and $u_n = t_n\phi_k + w_n$. Note that, as before, $w_n \to w^*$ uniformly in Ω. Moreover, setting

$$\Omega' := \Omega^+ \cup \Omega^-,$$

one has $u_n(x) = w_n(x)$ for all $x \in \Omega'$, and hence

$$t_n\Gamma(t_n) = \int_{\Omega'} b(u_n)u_n - \int_{\Omega'} b(u_n)w_n.$$

Furthermore, there results

$$|u_n(x)| = |t_n\phi_k(x) + w_n(x)| \to \infty \text{ for } |t_n| \to \infty \text{ and } x \notin \Omega'.$$

Using this and (b5) we deduce

$$\int_{\Omega'} b(u_n)u_n \to \sigma|\Omega'|. \tag{1.14}$$

Moreover, since $b(s) \to 0$ as $|s| \to \infty$ we also find

$$\int_{\Omega'} b(u_n)w_n \to 0. \tag{1.15}$$

Lastly (1.14)–(1.15) yield

$$t_n\Gamma(t_n) \to \sigma|\Omega'|.$$

Since, plainly, $|\Omega'| > 0$, the preceding limit is positive and the equation $\Gamma(t) = 0$ has a solution.

Remark 1.11 The same arguments show that there exists $\varepsilon > 0$ (depending on Ph) such that (PR) has at least two solutions provided

$0 < |(h|\phi_k)| < \varepsilon$. In addition, in a way similar to that sketched in Remark 1.9 (i), the same existence and multiplicity result can be proved relaxing (b3) and the assumption that λ_k is simple; see [**AM2**].

We end this section with a uniqueness result.

Theorem 1.12 *Consider* (PR) *with* $\lambda_k = \lambda_1$ *and suppose* b *satisfies* (b1), (b3) *with* $k = 1$, (b4) *and is such that* $b'(s) \neq 0$ *for all* s *(for example, let us take* $b'(s) > 0$).
  Then (PR) *has a unique solution if and only if*

$$b^- \int_\Omega \phi_1 < \int_\Omega h\phi_1 < b^+ \int_\Omega \phi_1. \qquad (1.16)$$

Proof. Note that (1.16) is nothing but (1.4): in fact now $\Omega = \Omega^+$ and $\Omega^- = \emptyset$. We will prove the theorem showing that Γ is strictly increasing. For this, let us recall that w is differentiable with respect to t (Lemma 1.7) with derivative denoted by w'. From

$$\Delta w + \lambda_1 w + Pb(t\phi_1 + w) = Ph$$

it follows that w' satisfies

$$\Delta w' + \lambda_1 w' + Pb'(t\phi_1 + w)(\phi_1 + w') = 0. \qquad (1.17)$$

As for Γ, one has that Γ is differentiable with derivative

$$\Gamma'(t) = \int_\Omega b'(t\phi_1 + w(t))(\phi_1 + w'(t))\phi_1.$$

Then (1.17) becomes

$$\Delta w' + \lambda_1 w' + b'(t\phi_1 + w)(\phi_1 + w') - \Gamma'(t)\phi_1 = 0. \qquad (1.18)$$

Suppose the contrary, that $\exists t^*$ such that $\Gamma'(t^*) = 0$. Then, setting $u^* = t^*\phi_1 + w(t^*)$ and $z^* = \phi_1 + w'(t^*)$, from (1.18) we infer

$$\Delta z^* + \lambda_1 z^* + b'(u^*)z^* = 0.$$

This means that z^* is a solution of the linear b.v.p. $\Delta z^* + m^* z^* = 0$, where $m^* = \lambda_1 + b'(u^*)$. Notice that z^* is not identically zero (in fact $(z^*|\phi_1) = (\phi_1|\phi_1) = 1$) and therefore $\lambda_k(m^*) = 1$ for some integer $k \geq 1$. By assumption $m^* < \lambda_2$ and hence the comparison property of eigenvalues (see Theorem 0.6 (ii)) yields $\lambda_2(m^*) > 1$. Thus one must have $\lambda_1(m^*) = 1$ and z^* is either > 0 or < 0 in Ω. Since $b' > 0$ and $\phi_1 > 0$, it follows that $\Gamma'(t^*) = \int_\Omega b'(u^*)z^*\phi_1$ is > 0 (resp. < 0) according to the sign of z^*. This is a contradiction, proving the theorem.

Remark 1.13 All the preceding results could be easily extended to elliptic problems like

$$\left.\begin{array}{l} \mathcal{L}u + au + b(u) = h, \\ u|_{\partial\Omega} = 0, \end{array}\right\}$$

where $\mathcal{L}$ is an elliptic operator with smooth coefficients (see subsection 0.6) and b such that $b(s)/s \to 0$ as $|s| \to \infty$. Moreover, h could be taken in $L^2(\Omega)$: in such a case one should work in Sobolev spaces instead of Hölder spaces and one would find weak solutions.

2 Problems with asymmetric nonlinearities

In this section we consider problem (D) in the case when $p(s)$ has two different asymptotes as $s \to \pm\infty$. More precisely we start by assuming that p satisfies

(p1) $p \in C^2(\mathbb{R}), p(0) = 0$ and $p''(s) > 0$ for all $u \in \mathbb{R}$,
(p2) $p'(s) \to \gamma'$ (resp. γ'') as $s \to -\infty$ (resp. $+\infty$), and there results

$$0 < \gamma' < \lambda_1 < \gamma'' < \lambda_2.$$

We keep the same notation as in the preceding section, letting $X = \{u \in C^{2,\alpha}(\overline{\Omega}) : u = 0 \text{ on } \partial\Omega\}$, $Y = C^{0,\alpha}(\overline{\Omega})$ and $F(u) = \Delta u + p(u)$. Note that (p1) implies that $F \in C^2(X, Y)$. The main abstract tool will be Theorem 3.2.6.

We start by proving the following.

Lemma 2.1 *F is proper.*

Proof. Let $u_n \in X$ be such that $F(u_n) = h_n$ is bounded in Y. We claim that $\|u_n\|_Y$ is bounded. Since the arguments are very close to those of Lemma 1.1, we will indicate the main changes only. Let supposing the contrary, $\|u_n\|_Y \to \infty$; then letting $z_n = u_n/\|u_n\|_Y$ one has

$$\Delta z_n + \varphi(u_n)z_n = \frac{h_n}{\|u_n\|_Y} \tag{2.1}$$

where

$$\varphi(u) = \begin{cases} p(u)/u & \text{if } u \neq 0, \\ p'(0) & \text{if } u = 0. \end{cases}$$

Since φ is bounded, (2.1) and Theorem 0.5 (ii) imply that

$$\|z_n\|_{C^{1,\alpha}} \leq \text{const.}$$

and, up to a sub-sequence, $z_n \to z^*$ in $C^1(\Omega)$ with $\|z^*\|_Y = 1$. Multiplying (2.1) by $w \in C_0^\infty(\Omega)$ and integrating we find

$$-\int_\Omega \nabla w \nabla z_n + \int_\Omega w\varphi(u_n)z_n = \int_\Omega \frac{wh_n}{\|u_n\|_Y}. \tag{2.2}$$

Note that if $z^*(x) < 0$ (resp. > 0) then $u_n(x) \to -\infty$ (resp. $+\infty$).
Thus, letting

$$m(x) = \begin{cases} \gamma' & \text{if } z^*(x) < 0, \\ \gamma'' & \text{if } z^*(x) > 0, \\ p'(0) & \text{if } z^*(x) = 0, \end{cases}$$

one has that $\varphi(u_n(x))z_n(x) \to m(x)z^*(x)$ pointwise in Ω. By the
Lebesgue Dominated Convergence Theorem, we infer from (2.2)

$$-\int_\Omega \nabla w \nabla z^* + \int_\Omega mwz^* = 0, \text{ for all } w \in C_0^\infty(\Omega).$$

Therefore $\Delta z^* + mz^* = 0$ and $\lambda_k(m) = 1$ for some integer $k \geq 1$.
Since $m \leq \gamma'' < \lambda_2$ it follows from Theorem 0.6 (ii) that $\lambda_2(m) > 1$
and hence one has $\lambda_1(m) = 1$. As a consequence z^* does not change
sign in Ω. Then m equals either γ' or γ'' and in both cases we reach
a contradiction. This proves the claim. This *a priori* estimate implies
readily that F is proper as shown in Corollary 1.3.

Next we want to study the singular set

$$\Sigma' = \{u \in X : F'(u) \notin \text{Inv}(X, Y)\}$$

. Notice that here $F'(u)$ is nothing but the map $v \to \Delta v + p'(u)v$ and
hence $u \in \Sigma'$ whenever the linear b.v.p.

$$\left. \begin{array}{l} \Delta v + p'(u)v = 0 \text{ in } \Omega, \\ u = 0 \text{ on } \partial\Omega, \end{array} \right\} \tag{2.3}$$

has solutions other than the trivial one, or in other words, whenever
$\lambda_k(p'(u)) = 1$ for some integer $k \geq 1$. Since, by assumption, $p'(s) <
\gamma'' < \lambda_2$, the comparison property (Theorem 0.6 (ii)) implies that 1 is ,
if an eigenvalue, the *first* eigenvalue of (2.3).

Lemma 2.2

(i) Σ' *is not empty, closed and connected.*

(ii)   *Every $u \in \Sigma'$ is an ordinary singular point.*

Proof. To prove (i) we will show that Σ' has a Cartesian representation
on a linear subspace W of X of codimension 1. Fixing $z \in X, z(x) > 0$
for all $x \in \Omega$, let W be any linear subspace of X such that $z \notin W$. Every

$u \in X$ can be written in the form $u = \sigma z + w$, with $\sigma \in \mathbb{R}$ and $w \in W$. Let $m_\sigma = p'(\sigma z + w)$ and consider the first eigenvalue $\lambda_1(m_\sigma)$ of

$$\left.\begin{array}{r} \Delta v + \lambda m_\sigma v = 0 \text{ in } \Omega, \\ u = 0 \text{ on } \partial\Omega. \end{array}\right\}$$

According to the preceding remarks, the point $\sigma z + w$ belongs to the singular set Σ' if and only if $\lambda_1(m_\sigma) = 1$. Since $z(x) > 0$, $m_\sigma > m_\mu$ whenever $\sigma > \mu$ and hence $\lambda_1(m_\sigma)$ is a decreasing function of σ (see Theorem 0.6 (ii)). Moreover, from $z(x) > 0$ in Ω, we also infer that $m_\sigma \to \gamma'(\gamma'')$ as $\sigma \to -\infty$ (resp. $+\infty$), pointwise and in L_q (for all q) because p' is bounded. Using the continuity property of the eigenvalues (Theorem 0.6 (iv)) we deduce that

$$\left.\begin{array}{l} \lambda_1(m_\sigma) \to \lambda_1(\gamma') = \dfrac{\lambda_1}{\gamma'} > 1 \text{ as } \sigma \to -\infty, \\[2mm] \lambda_1(m_\sigma) \to \lambda_1(\gamma'') = \dfrac{\lambda_1}{\gamma''} < 1 \text{ as } \sigma \to +\infty. \end{array}\right\} \tag{2.4}$$

Since $\lambda_1(m_\sigma)$ is decreasing, it follows from (2.4) that there is a unique σ^* such that $\lambda_1(m_{\sigma^*}) = 1$. Since the corresponding $u^* = \sigma^* z + w$ belongs to Σ', (i) follows.

(ii) Let $u \in \Sigma'$. Then $\lambda_1(p'(u)) = 1$. Since the first eigenvalue is simple, Condition (a) of Section 3.2 holds. In particular, $\operatorname{Ker}(F'(u))$ is spanned by a non-zero function ϕ, which does not change sign on Ω; moreover, $h \in R(F'(u))$ if and only if $\int_\Omega \phi h = 0$ (see Theorem 0.7 (ii)) so that $R(F'(u)) = \operatorname{Ker}(\psi)$, where

$$\psi : h \to \int_\Omega \phi h.$$

To prove that Condition (b) of Section 3.2 holds, we notice that $F''(u)[v, w] = p''(u)vw$. Then we have

$$\langle \psi, F''(u)[\phi, \phi] \rangle = \int_\Omega p''(u)\phi^3,$$

which does not vanish because $p''(s) > 0$ (Hypothesis (p1)) and ϕ is either positive or negative on Ω. This proves that $\langle \psi, F''(u)[\phi, \phi] \rangle \neq 0$ and hence that u is an ordinary singular point.

Lemma 2.3 *For all $h \in F(\Sigma')$, the equation $F(u) = h$ has a unique solution.*

Proof. Let $z \in \Sigma'$ be such that $h = F(z)$. Suppose the contrary that there is $w \neq z$ with $F(w) = h$. Set

$$\omega(x) = \begin{cases} (p(w(x)) - p(z(x)))/(w(x) - z(x)) & \text{for all } x : w(x) \neq z(x), \\ p'(z(x)) & \text{for all } x : w(x) = z(x). \end{cases}$$

From $F(z) = F(w)$ (that is , $\Delta z + p(z) = \Delta w + p(w)$) it follows immediately that $v := w - z$ is a (non-trivial) solution of

$$\left. \begin{array}{r} \Delta v + \omega v = 0 \text{ in } \Omega, \\ v = 0 \text{ on } \partial\Omega. \end{array} \right\}$$

In other words, there results $\lambda_k(\omega) = 1$, for some integer $k \geq 1$. Since $\gamma' < \omega < \gamma'' < \lambda_2$, then we must have $\lambda_1(\omega) = 1$ and $v(= w - z)$ does not change sign on Ω. Suppose, for example, that $v > 0$, that is, $w > z$ (if $v < 0$ the argument is similar). Since $p'' > 0$, $\omega(x) < p'(z(x))$ on Ω and hence $(1 =)\lambda_1(\omega) > \lambda_1(p'(z))$. But $z \in \Sigma'$ implies that $\lambda_1(p'(z)) = 1$, a contradiction.

Lemmas 2.1, 2.2 and 2.3 allow us to apply Theorem 3.2.6. yielding a precise description concerning the solutions of (D).

Theorem 2.4 *Let p satisfy* (p1–2). *Then $Y = C^{0,\alpha}(\overline{\Omega}) = Y_0 \cup Y_1 \cup Y_2$ with the following properties:*

(1) *Y_1 is a C^1-manifold of codimension 1 in Y and (D) has a unique solution for all $h \in Y_1$;*

(2) *Y_0 and Y_2 are disjoint open subsets of Y such that for all $h \in Y_2$ (D) has exactly two solutions, while for all $h \in Y_0$ problem (D) has no solution.*

Theorem 2.4 is taken from [**AP**]. A different proof has been given by Berger & Podolak [**BP**] by using arguments more on the line of those discussed in the preceding section for problems at resonance. Roughly, a Lyapunov–Schmidt procedure yields the reduction of problem (D) to the study of a one-dimensional equation $\Gamma(t) = (h, \phi_1)$; moreover the properties of p allow one to show that $\Gamma(t) \to +\infty$ as $|t| \to +\infty$ and that $\Gamma''(t) > 0$ for all t.

Problems like the preceding have been called problems with "jumping nonlinearities" to point out the fact that the nonlinearity p has two different asymptotes at $+\infty$ and at $-\infty$.

Also in the case of problems with "jumping nonlinearities" the *existence* (and multiplicity) results could be obtain by different methods and could be improved. Below we will expound a rather general result due to Kazdan & Warner [**KW**]. First some preliminaries are in order. To highlight the different behaviour of p at $\pm\infty$, we will take p of the

form
$$p(s) = \beta s^+ - \alpha s^- + b(s),$$
where $\alpha, \beta \in \mathbb{R}$, $\beta \neq \alpha$ (we will take below $\alpha < \beta$), $s^+ = \max(s, 0)$, $s^- = s^+ - s$, and b is bounded. Note that the nonlinearity p in Theorem 2.4 is of this form with $\beta = \gamma''$ and $\alpha = \gamma'$.

It is also convenient to set
$$-h(x) = t\phi(x) + \eta(x), \text{ with } (\eta|\phi) = 0,$$
where $\phi = \phi_1 > 0$, $b(x, s) = b(s) + \eta(x)$ and $p(x, s) = \beta s^+ - \alpha s^- + b(x, s)$. With this notation (D) becomes
$$-\Delta u = p(x, u) + t\phi \text{ in } \Omega, \ u = 0 \text{ on } \partial\Omega. \tag{P_t}$$
The index t is used to highlight that the results will depend on t.

First of all, let us note that if the interval $[\alpha, \beta]$ does not contain any eigenvalue λ_k, then (P_t) has solutions for all $t \in \mathbb{R}$. This could be proved by topological degree arguments, but will not be discussed here.

The situation is different when some $\lambda_k \in (\alpha, \beta)$, a case in which non-existence or multiplicity results can arise. More precisely, we prove the following.

Theorem 2.5 *Suppose that*
(p3) *b is Hölder-continuous and $|b(x, s)| \leq M$, for all $(x, s) \in \Omega \times \mathbb{R}$,*
(p4) *$-\infty < \alpha < \lambda_1 < \beta < +\infty$.*
Then there exists $T^* \in \mathbb{R}$ such that
(i)   *for all $t \leq T^*$ problem (P_t) has a solution,*
(ii)   *for all $t > T^*$ problem (P_t) has no solution.*

The proof of Theorem 2.5 will be carried out using sub- and super-solutions. Recall that $\chi \in C^2(\Omega)$ is a sub-solution of the problem
$$\left.\begin{array}{r}-\Delta u = g(x, u) \text{ in } \Omega, \\ u = 0 \text{ on } \partial\Omega, \end{array}\right\} \tag{2.5}$$
if there results
$$\left.\begin{array}{r}-\Delta\chi \leq g(x, \chi) \text{ in } \Omega, \\ \chi \leq 0 \text{ on } \partial\Omega. \end{array}\right\} \tag{2.6}$$
A super-solution ψ is defined similarly, reversing the inequalities in (2.6).

It is well known (see for example [**Ama**]) that the following result holds.

Proposition 2.6 *Let p be continuous in $\Omega \times \mathbb{R}$ and suppose that (2.5) has a sub-solution χ and a super-solution ψ such that $\chi \leq \psi$ on Ω. Then (2.5) has a solution u with $\chi \leq u \leq \psi$.*

Let us start by noticing that from $\alpha < \beta$ and $|b| \leq M$ it follows that

$$p(x, u) \geq \alpha u^+ - \alpha u^- - M = \alpha u - M. \tag{2.7}$$

For any $t \in \mathbb{R}$, consider the boundary-value problem

$$-\Delta u = \alpha u - M + t\,\phi,\ u_{|\partial\Omega} = 0. \tag{2.8}$$

Since $\alpha < \lambda_1$, (2.8) has a unique solution χ_t. Using (2.7) it follows immediately that

$$-\Delta\chi_t = \alpha\chi_t - M + t\phi \leq p(x, \chi_t) + t\,\phi.$$

Therefore, we can say that the following lemma holds

Lemma 2.7 *For all $t \in \mathbb{R}$, χ_t is a sub-solution for* (P$_t$).

Next we prove another lemma.

Lemma 2.8 *If ψ is any super-solution of* (P$_t$) *then $\psi \geq \chi_t$.*

Proof. Since ψ is a super-solution of (P$_t$) and χ_t solves (2.8) one infers
$$-\Delta(\psi - \chi_t) \geq p(x, \psi) + t\phi + \Delta\chi_t$$
$$= p(x, \psi) + t\phi - (\alpha\chi_t - M + t\phi)$$
$$= p(x, \psi) + M - \alpha\chi_t.$$
Using (2.7) we deduce that

$$-\Delta(\psi - \chi_t) \geq \alpha(\psi - \chi_t) \text{ in } \Omega.$$

Since $\alpha < \lambda_1$ and $\psi - \chi_t \geq 0$ on $\partial\Omega$ the Maximum Principle 0.8 applies to $w = \psi - \chi_t$ yielding $\psi - \chi_t \geq 0$ in Ω, as required.

The next step is to prove a third lemma.

Lemma 2.9 *There exists $t^* \in \mathbb{R}$ such that for all $t \leq t^*$,* (P$_t$) *has a super-solution ψ_t.*

Proof. Let us fix $\sigma > 0$ and take $m > 0$ satisfying

$$m \geq \max\{p(x, s) : x \in \Omega,\ 0 \leq s \leq \sigma\}.$$

Moreover, let $\Omega' \subset\subset \Omega$ (i.e. Ω' has closure contained in Ω) be such that $|\Omega'| < \varepsilon$, where $\varepsilon > 0$ is a number to be determined later on, and set $\Omega^* = \Omega\backslash\Omega'$. Fixing $\Omega'' \subset\subset \Omega'$, consider a function $h \in C^\infty(\Omega)$, such that (i) $0 \leq h \leq m$ in Ω, (ii) $h(x) = 0$ in Ω'', (iii) $h(x) = m$ in Ω^* and let us denote by ψ the solution of $\Delta\psi = h$, $\psi_{|\partial\Omega} = 0$. From the Maximum Principle (Theorem 0.8) $\psi \geq 0$ in Ω. Moreover from Theorem 0.5(i) one has

$$\|\psi\|_{H^{2,p}} \leq c_1\|h\|_{L^p} \leq c_1\varepsilon^{1/p} \text{ for all } p > 1.$$

Taking $p > n/2$ (where n is the dimension of Ω), we deduce, using Theorem 0.4(iii),

$$\|\psi\|_\infty \leq c_2 \varepsilon^{1/p}.$$

Therefore, if ε is taken in such a way that $c_2 \varepsilon^{1/p} \leq \sigma$, then one has $0 \leq \psi(x) \leq \sigma$ and hence

$$p(x, \psi(x)) \leq m. \tag{2.9}$$

Let $\phi_0 = \min\{\phi(x) : x \in \Omega'\}(> 0)$ and set $t^* = -m/\phi_0$. We claim that

$$t^*\phi + m \leq h \text{ in } \Omega \tag{2.10}$$

Indeed, if $x \in \Omega'$ then

$$t^*\phi + m = -\frac{m}{\phi_0}\phi + m \leq 0 \leq h,$$

while if $x \in \Omega^*$ one has $t^*\phi + m < m = h$, and (2.10) follows.

Using this and (2.9) and (2.10) we find

$$-\Delta\psi = h \geq t^*\phi + m \geq p(x, \psi) + t^*\phi.$$

Lastly, for all $t \leq t^*$ one has

$$-\Delta\psi \geq p(x, \psi) + t^*\phi \geq p(x, \psi) + t\phi,$$

proving the lemma.

Proof of Theorem 2.5 Define T^* by setting

$$T^* = \sup\{t \in \mathbb{R} : (P_t) \text{ has a super-solution}\}.$$

From Lemma 2.9 it follows that T^* is well defined. Moreover it is easy to see that $T^* < +\infty$. In fact, if (P_t) has a super-solution w, that is $-\Delta w \geq p(x, w) + t\phi$, then multiplying by ϕ and integrating on Ω we find

$$\lambda_1 \int_\Omega \phi w \geq \int_\Omega \phi p(x, w) + t$$

$$= \int_\Omega \phi[\beta w^+ - \alpha w^-] + \int_\Omega \phi b(x, w) + t.$$

Hence

$$t \leq \lambda_1 \int_\Omega \phi w - \int_\Omega \phi[\beta w^+ - \alpha w^-] + c,$$

where $c = M \int_\Omega \phi$. Since $\lambda_1 w + \alpha w^- - \beta w^+ = \lambda_1(w^+ - w^-) + \alpha w^- - \beta w^+ < 0$ because $\alpha < \lambda_1 < \beta$. Then it follows that $t \leq c$ and $T^* < +\infty$.

For all $t < T^*$ (P_t) has a super-solution ψ_t and a sub-solution χ_t (Lemma 2.7); moreover from Lemma 2.8 it follows that $\chi_t \leq \psi_t$. Using Proposition 2.6 we infer that for all $t < T^*$ (P_t) has a solution u with

$\chi_t \leq u \leq \psi_t$. It remains to show that (P_t) has a solution for $t = T^*$. Take a sequence $t_k \to T^*, t_k < T^*$. Problems (P_k) corresponding to values $t = t_k$ have solutions u_k. From the preceding construction it is easy to check that the u_k converge to some u^* which solves (P_{T^*}).

This completes the proof of Theorem 2.5.

Remarks 2.10

(i) Improving Theorem 2.5, it has been shown [**AmaH**] that for all $t < T^*(P_t)$ has at least two distinct solutions (the second one is found by degree-theoretic arguments). See Problems (1) and (2) below for another multiplicity result in this direction. For an extensive discussion of elliptic equations with jumping nonlinearities we refer to [**Fu**].

(ii) A geometric description of the range of a differential operator more in the spirit of Theorem 2.4 can be found in [**McS**].

5

Bifurcation results

The structure of the solution set of a nonlinear functional equation can be very complicated and often it could be convenient to assume a "genetic" point of view, seeking for when new solutions are generated, near a given one, after a small perturbation. A convenient device consists in finding (or introducing) a parameter λ, and studying an equation $F(\lambda, u) = 0$ which possesses a fixed solution for all values of the parameter. An interesting phenomenon is when there is a "branching" of new solutions of $F(\lambda, u) = 0$ in correspondence with some value of the parameter. This is the object of the "Bifurcation theory" we will discuss in this chapter in its more elementary aspects.

1 Introduction

Let X, Y be two Banach spaces. We are interested in studying equations of the type

$$F(\lambda, u) = 0 \qquad (1.1)$$

where

$$F : \mathbb{R} \times X \to Y$$

is a map depending on a *real* parameter λ.

As we will see in the next chapter, equations like (1.1) model a broad class of problems arising in applications, where the parameter λ often has a physical interpretation: it can be the intensity of the loading in some elasticity problems, the Rayleigh number in hydrodynamics, and so on.

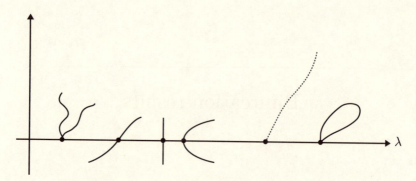

Figure 5.1 This and the following bifurcation diagrams
are to be interpreted as suggestions only.

In this chapter we will always assume that $F \in C^2(\mathbb{R} \times X, Y)$ and
that

$$F(\lambda, 0) = 0 \text{ for all } \lambda \in \mathbb{R}.$$

If this is true, then (1.1) has for all λ the solution $u = 0$, which will
be referred to as the *trivial solution*.
  $\mathcal{S} = \{(\lambda, u) \in \mathbb{R} \times X : u \neq 0, F(\lambda, u) = 0\}$ will denote the set of
non-trivial solutions of (1.1).

It can happen that for some values of the parameter there are one
or more solutions of (1.1) that branch off from the trivial one. These
values of λ are called the *bifurcation points* of (1.1) (Figure 5.1). More
precisely, we give the following definition:

Definition 1.1   We say that λ^* is a *bifurcation point* for F (from the
trivial solution) if there is a sequence $(\lambda_n, u_n) \in \mathbb{R} \times X$ with $u_n \neq 0$ and
$F(\lambda_n, u_n) = 0$ such that

$$(\lambda_n, u_n) \to (\lambda^*, 0).$$

Another, equivalent, way to define a bifurcation point, is to require
that $(\lambda^*, 0)$ belong to the closure (in $\mathbb{R} \times X$) of $\mathcal{S}$, that is, that in any
neighbourhood of $(\lambda^*, 0)$ there is a point $(\lambda, u) \in \mathcal{S}$.

Let us begin our discussion by stating a result that follows immediately
from the Implicit Function Theorem.

Proposition 1.2   *A necessary condition for λ^* to be a bifurcation point
for F is that the partial derivative $F_u(\lambda^*, 0)$ is not invertible.*

Proof. If $F_u(\lambda^*, 0) \in \text{Inv}(X, Y)$ then Theorem 2.2.3 applies and there exists a neighbourhood $\Theta \times V$ of $(\lambda^*, 0)$ such that

$$F(\lambda, u) = 0, (\lambda, u) \in \Theta \times V, \Longleftrightarrow u = 0.$$

Therefore λ^* is not a bifurcation point for F.

An interesting case is when $X = Y$ and

$$F(\lambda, u) = \lambda u - G(u). \tag{1.2}$$

In such a case $F_u(\lambda^*, 0) = \lambda^* I - G'(0)$ and Proposition 1.2 becomes the following.

Proposition 1.3 *If λ is a bifurcation point for F of the form (1.2) then λ belongs to the spectrum $\sigma[G'(0)]$ of $G'(0)$.*

It is quite natural to ask whether or not Proposition 1.3 can be inverted:

if $\lambda \in \sigma[G'(0)]$, is λ a bifurcation point for F ?

We anticipate that, in this generality, the answer to the preceding question is negative.

The particular case when F has the form (1.2) with $G = A \in L(X)$ is particularly enlightening. Note that, if G is linear, then $G'(0) = A$ and the relationships between the bifurcation points for

$$F = \lambda I - A \tag{1.3}$$

and the spectrum $\sigma(A)$ of A can be established in a precise fashion.

First of all, it is clear that the eigenvalues of A are bifurcation points for $F = (1.3)$.

Moreover, the following result can readily be proved.

Proposition 1.4 *Let $A \in L(X)$ and $F(\lambda, u) = \lambda u - A(u)$. Then λ^* is a bifurcation point for F if and only if λ^* belongs to the closure of the eigenvalues of A.*

Remarks 1.5

(i) As a consequence of Proposition 1.4, we deduce that *in general, there might be points λ belonging to the spectrum of A that are not bifurcation points for F of the form (1.3).*

(ii) From Proposition 1.4 it also follows that λ^* can be a bifurcation point for $\lambda I - A$ without being an eigenvalue of A.

We have seen that when F has the form (1.3) all the eigenvalues of A are bifurcation points for F. The following example shows that in the nonlinear case a value λ^* can be an eigenvalue of $G'(0)$ without being a bifurcation for $F(\lambda, u) = \lambda u - G(u)$.

Example 1.6 Let $X = Y = \mathbb{R}^2$, and consider the application $G : X \to X$ defined by

$$G(x,y) = (x + y^3, y - x^3).$$

The value $\lambda^* = 1$ is an eigenvalue of $g'(0) = I$, but it is not a bifurcation for

$$F : (x,y) \to \lambda(x,y) - G(x,y).$$

For let (x,y) be a solution of $F = 0$. From

$$\left. \begin{array}{l} \lambda x = x + y^3, \\ \lambda y = y - x^3, \end{array} \right\}$$

it follows that

$$x^4 + y^4 = 0,$$

and hence $(x,y) = (0,0)$. Therefore $F = 0$ has only the trivial solution and there are no bifurcation points for F.

2 Some elementary examples

In this section we will discuss a couple of simple examples, where the existence of bifurcation points can be proved in a rather elementary way.

As a first example, let us consider the boundary-value problem

$$\frac{d^2u}{dt^2} + \lambda(u - u^3) = 0, \quad t \in [0, \pi], \tag{2.1}$$

$$u(0) = u(\pi) = 0. \tag{2.1'}$$

For all values of the parameter $\lambda \in \mathbb{R}$ (2.1)–(2.1') has the *trivial solution* $u(t) \equiv 0$. To find other possible solutions we can work in the phase plane and argue as follows. Multiplying (2.1) by

$$p = \frac{du}{dt}$$

one finds immediately that any solution of (2.1) satisfies the energy relationship

$$\frac{1}{2}p^2 + \lambda(\frac{u^2}{2} - \frac{u^4}{4}) = c \ (= \text{constant}). \tag{2.2}$$

The integral curves in the phase plane (u, p) are represented in Figure 5.2.

We can distinguish between two families of integrals, which are separated by curves that pass through the singular points $(1, 0)$ and $(-1, 0)$. In particular, the closed curves correspond to periodic solutions of (2.1) and we are interested in the arcs of those closed integrals that start from

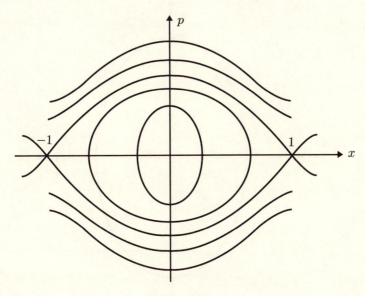

Figure 5.2

points $(0, p)$ and again reach the p-axis after a time equal to π. From the symmetry, such a time is an integer multiple of the semiperiod T.

Let Γ_ξ be an integral curve crossing the u-axis at $u = \xi$, with $0 < \xi < 1$ (see Figure 5.3).

Putting $p = 0$ and $u = \xi$ in (2.2) we can find the value of c corresponding to Γ_ξ:

$$c = \lambda(\frac{1}{2}\xi^2 - \frac{1}{4}\xi^4). \qquad (2.3)$$

Let $T(\lambda, \xi)$ be the semiperiod of Γ_ξ. From the symmetry, $T(\lambda, \xi)$ is given by

$$T(\lambda, \xi) = 2\int_\gamma \frac{du}{p},$$

where γ denotes the arc of Γ_ξ with $p \geq 0$.

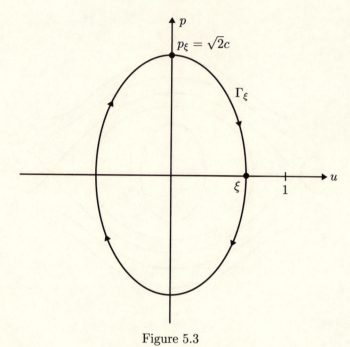

Figure 5.3

Taking into account (2.2) and (2.3) one finds with easy calculations

$$T(\lambda,\xi) = 2 \int_0^\xi \frac{dx}{\sqrt{[2c - \lambda(x^2 - \frac{1}{2}x^4)]}}$$

$$= 2 \int_0^1 \frac{\xi\,dy}{\sqrt{[2c - \lambda(\xi^2 y^2 - \frac{1}{2}\xi^4 y^4)]}}$$

$$= 2 \int_0^1 \frac{\xi\,dy}{\sqrt{[2\lambda(\frac{1}{2}\xi^2 - \frac{1}{4}\xi^4) - \lambda(\xi^2 y^2 - \frac{1}{2}\xi^4 y^4)]}} \tag{2.4}$$

$$= \frac{2}{\sqrt{\lambda}} \int_0^1 \frac{dy}{\sqrt{[1 - \frac{1}{2}\xi^2 - y^2(1 - \frac{1}{2}\xi^2 y^2)]}}.$$

Let us note explicitly that (2.4) allows us to extend $T(\lambda,.)$ at $\xi = 0$ by setting

$$T(\lambda,0) = \frac{2}{\sqrt{\lambda}} \int_0^1 \frac{dy}{\sqrt{(1 - y^2)}} = \frac{\pi}{\sqrt{\lambda}}. \tag{2.5}$$

As anticipated before, Γ_ξ gives rise to a solution of the boundary-value

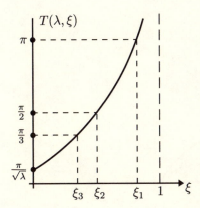

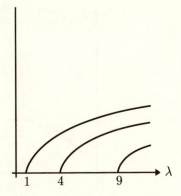

Figure 5.4

problem (2.1)–(2.1′) whenever ξ is such that

$$mT(\lambda, \xi) = \pi \qquad (2.6)$$

for some $m \in \mathbb{N}$.

For example, if ξ is such that $T(\lambda, \xi) = \pi$, then Γ_ξ corresponds to a positive solution of (2.1) with $u(0) = 0$ and $p(0) = u'(0) = p_\xi$ (see Fig.5.3), $u(\pi) = 0$ and $p(\pi) = u'(\pi) = -p_\xi$.

To discuss equation (2.6) we first deduce from (2.4) that, for any fixed λ, there result

$$\frac{\partial T(\lambda, \xi)}{\partial \xi} > 0,$$

$$T(\lambda, \xi) \to +\infty \text{ as } \xi \uparrow 1.$$

Taking (2.5) also into account, we get that

(a) if $\lambda < 1$ then $T(\lambda, \xi) > \pi$ for all ξ, and (2.1)–(2.1′) has only the trivial solution $u \equiv 0$,

(b) if $\lambda = 1$ then $T(\lambda, 0) = \pi$ and $\xi = 0$ is the only solution of (2.6), and hence (2.1)–(2.1′) has again the trivial solution only,

(c) if $1 < \lambda < 4$, then (2.6) (with $m = 1$) has a unique solution $\xi \neq 0$ which corresponds to a positive solution u_ξ of (2.1)–(2.1′). (More precisely, as $\lambda \downarrow 1$ the solution $\xi = \xi(\lambda)$ as well as p_ξ tends to 0; correspondingly (2.1)–(2.1′) has a family $u_\lambda = u_{\xi(\lambda)}$ of (positive) solutions, depending continuously on λ, such that $\|u_\lambda\|_{C^1} \to 0$ as $\lambda \downarrow 1$, and we can say that the (positive) solutions of (2.1)–(2.1′) bifurcate from the trivial solution at the value $\lambda = 1$ of the parameter).

The discussion can be carried over showing that (see Figure 5.4)

(d) if $k^2 < \lambda < (k+1)^2$ then (2.6) has k solutions $\xi_1, \ldots, \xi_k \neq 0$,

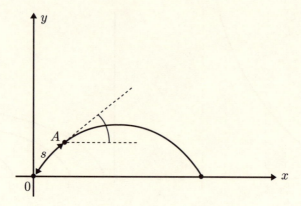

Figure 5.5

satisfying

$$h\,T(\lambda, \xi_h) = \pi \quad (h = 1, \ldots, k).$$

Each ξ_h corresponds to a solution of (2.1)–(2.1′) with precisely $h - 1$ nodes in $(0, \pi)$. In particular we can say that there is a continuous family of solutions u_λ of (2.1)–(2.1′), with $k - 1$ nodes in $(0, \pi)$, such that $\|u_\lambda\|_{C^1} \to 0$ as $\lambda \downarrow k^2$.

It is worth noting that, from the abstract point of view, the boundary-value problem (2.1)–(2.1′) gives rise to a functional equation of the type (1.1). Here X is the Banach space of $C^2(0, \pi)$ functions vanishing at $t = 0$ and $t = \pi$, $Y = C(0, \pi)$, and $F : \mathbb{R} \times X \to Y$ is given by (see also Example 2.1.5)

$$F(\lambda, u) = \frac{\mathrm{d}^2 u}{\mathrm{d}t^2} + \lambda(u - u^3).$$

There results

$$F_u(\lambda, 0) : u \to \frac{\mathrm{d}^2 u}{\mathrm{d}t^2} + \lambda u,$$

and the values $\lambda_k = k^2$ are precisely the eigenvalues of the linear problem

$$\frac{\mathrm{d}^2 u}{\mathrm{d}t^2} + \lambda u = 0, \ u(0) = u(\pi) = 0.$$

As a second example, we consider the buckling problem for an elastic beam of length L.

We suppose that one edge of the beam is hinged, while the other one is variable on the x-axis. The beam is compressed at the free edge by a force of intensity $K > 0$. Denote by $(x(s), y(s))$ the coordinates of a point A on the beam, as a function of the length s of the arc $0A$, and let $\phi(s)$ be the angle between the tangent to the beam at A and the x-axis. See Figure 5.5.

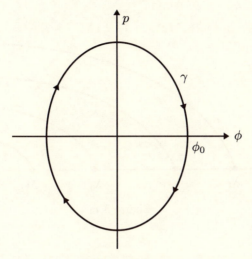

Figure 5.6

In accordance with the Euler–Bernoulli theory, the curvature of the beam at any point is proportional to the momentum of the applied force. Then there result

$$Ky = -\kappa \frac{d\phi}{ds} \quad (\kappa = \text{constant}), \quad \frac{dy}{ds} = \sin\phi, \tag{2.7}$$

together with the boundary conditions

$$y(0) = y(L) = 0. \tag{2.7'}$$

From (2.7)–(2.7') we deduce

$$\left. \begin{aligned} \frac{d^2\phi}{ds^2} + \lambda\sin\phi = 0, \quad s \in [0, L], \\ \phi'(0) = \phi'(L) = 0, \end{aligned} \right\} \tag{2.8}$$

where $\lambda = K/\kappa > 0$.

To study (2.8) we can proceed as before. We refer to Figure 5.6. If we let $p = d\phi/ds$, conservation of energy yields

$$\frac{1}{2}p^2 - \lambda\cos\phi = c = -\lambda\cos\phi_0,$$

and the semiperiod of the (closed) curve passing through $(\phi_0, 0)$ is given by

$$T(\lambda, \phi_0) = 2 \int_\gamma \frac{d\phi}{p}$$

$$= \frac{2}{\sqrt{\lambda}} \int_0^{\pi/2} \frac{d\theta}{\sqrt{(1 - \omega^2 \sin^2\theta)}},$$

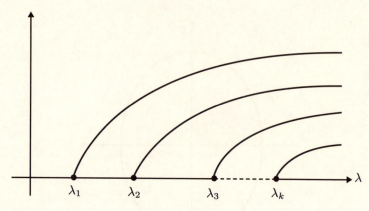

Figure 5.7

where

$$\omega = \sin \frac{\phi_0}{2}.$$

An arc of a closed curve joining two points of the x-axis corresponds to a solution of (2.8) whenever

$$hT(\lambda, \phi_0) = L,$$

for some $h \in \mathbf{N}$.

Since $T(\lambda, .)$ is still strictly increasing and $T(\lambda, \phi_0) \to +\infty$ as $\phi_0 \uparrow \pi$, we deduce:

(a) For $0 < \lambda \le \pi^2/L^2$ (2.8) has only the trivial solution $\phi \equiv 0$.

(b) For $k^2\pi^2/L^2 < \lambda \le (k+1)^2\pi^2/L^2$ (2.8) has k nontrivial solutions $\phi_1, \ldots, \phi_k$. In addition, for fixed $k = 1, 2, \ldots$, there exists a continuous family ϕ_λ of nontrivial solutions of (2.8) whose C^1-norm tends to zero as $\lambda \downarrow \lambda_k = k^2\pi^2/L^2$.

The bifurcation diagram is drawn in Figure 5.7.

As for the preceding case, we can see that the functional equation corresponding to (2.8) is

$$F(\lambda, \phi) := \frac{\mathrm{d}^2\phi}{\mathrm{d}s^2} + \lambda \sin \phi = 0$$

whith $\phi \in X$, the Banach space of functions of $C^2(0, L)$ such that $\phi'(0) = \phi'(L) = 0$.

The linearized equation $F_\phi(\lambda, 0)\psi = 0$ becomes

$$\frac{\mathrm{d}^2\psi}{\mathrm{d}s^2} + \lambda\psi = 0, \psi'(0) = \psi'(L) = 0, \tag{2.9}$$

whose positive eigenvalues are just

$$\lambda_k = \frac{k^2\pi^2}{L^2}, \quad k = 1, 2, \ldots.$$

Note that $\lambda = 0$ would also be a bifurcation point of $F(\lambda, \phi) = 0$, F given by (2.9). Indeed, for $\lambda = 0$, the equation $F(0, \phi) = 0$ has the family of nontrivial solutions $\phi =$ const. However, these solutions correspond to the trivial solution $y \equiv 0$ of the physical problem (2.7).

3 The Lyapunov–Schmidt reduction

In this section we dicuss a general procedure, introduced by Lyapunov [**Ly 1–2**] and Schmidt [**Schm**], which will be a basic tool hereafter, but can also be useful in several other situations. It is a method we have already used in Sections 3.2 and 4.1, for the specific problem studied there.

Let $F \in C^2(\mathbb{R} \times X, Y)$ be such that

$$F(\lambda, 0) = 0.$$

According to Proposition 1.2 the possible bifurcation points for F are the values λ^* such that $F_u(\lambda^*, 0)$ is not invertible. We set

$$L = F_u(\lambda^*, 0),$$
$$V = \mathrm{Ker}(L),$$
$$R = R(L),$$

and suppose that

(a)   *V has a topological complement W in X.*

This means that there exists a closed subspace W of X such that

$$X = V \oplus W \tag{3.1}$$

and any $u \in X$ can be written in the form

$$u = v + w, \; v \in V, \; w \in W. \tag{3.2}$$

On the range R of L we assume

(b)   *R is closed and has a topological complement Z in Y.*

This means that $Y = Z \oplus R$, with Z closed and such that $Z \cap R = \{0\}$. For example, (a) and (b) hold true when V is finite-dimensional and R has finite codimension, that is, when L is a *Fredholm operator*.

Next, let P and Q denote the conjugate projections onto Z and R, respectively.

Using (3.2) and applying P and Q one finds that $F(\lambda, u) = 0$ is equivalent to the system

$$PF(\lambda, v + w) = 0, \tag{3.3$'$}$$

$$QF(\lambda, v + w) = 0. \tag{3.3$''$}$$

For later use, it is also convenient to set

$$F(\lambda, u) = Lu + \varphi(\lambda, u).$$

Using (3.2) and recalling that $Lv = 0$, one has

$$F(\lambda, u) = Lw + \varphi(\lambda, v + w).$$

Recalling that $Lw \in R$, we get $QLw = Lw$. Then (3.3″) becomes

$$Lw + Q\varphi(\lambda, v + w) = 0. \tag{3.4}$$

We set

$$\Phi(\lambda, v, w) = Lw + Q\varphi(\lambda, v + w)$$

and note that $\Phi \in C^2(\mathbb{R} \times V \times W, R)$. Moreover

$$\Phi_w(\lambda^*, 0, 0) : w \to Lw + Q\varphi_u(\lambda^*, 0)w.$$

Since, by definition, $\varphi(\lambda, u) = F(\lambda, u) - Lu$, there results

$$\varphi_u(\lambda^*, 0) = F_u(\lambda^*, 0) - L = 0. \tag{3.5}$$

In other words, $\varphi_u(\lambda^*, 0)$ is the zero mapping in $L(X, Y)$ and therefore it follows that $\Phi_w(\lambda^*, 0, 0) = L|_W$.

We remark that the restriction $L|_W$ of L to W, as a map from W to R, is injective and surjective, Since R is closed, $(L|_W)^{-1}$ is continuous from R to W, namely

$$L|_W \in \text{Iso}\ (W, R). \tag{3.6}.$$

Hence $\Phi_w(\lambda^*, 0, 0) \in \text{Iso}\ (W, R)$, the Implicit Function Theorem applies to Φ and (3.4) can be uniquely solved, locally, with respect to w. To be precise, there exist

(i)   a neighbourhood Λ of λ^*,
(ii)   a neighbourhood $\mathcal{V}$ of $v = 0$ in V,
(iii) a neighbourhood $\mathcal{W}$ of $w = 0$ in W, and
(iv)   a function $\gamma \in C^2(\Lambda \times \mathcal{V}, \mathcal{W})$,

such that the unique solutions of (3.5″) in $\Lambda \times \mathcal{V} \times \mathcal{W}$ are given by $(\lambda, v, \gamma(\lambda, v))$.

In particular, for future reference, we remark that results

$$\gamma(\lambda, 0) = 0 \quad \text{for all } \lambda \in \Lambda. \tag{3.7}$$

Moreover one has

$$\gamma_v(\lambda^*, 0) = 0. \tag{3.8}$$

To see this, we can use the Implicit Function Theorem or else we can take into account that

$$L\gamma(\lambda, v) + Q\varphi(\lambda, v + \gamma(\lambda, v)) = 0 \quad \text{for all } (\lambda, v) \in \Lambda \times \mathcal{V}.$$

Differentiating with respect to v at $(\lambda^*, 0)$, and letting $\Gamma = \gamma_v(\lambda^*, 0)$, we find

$$L\Gamma x + Q\varphi_u(\lambda^*, \gamma(\lambda^*, 0))[x + \Gamma x] = 0 \text{ for all } x \in V.$$

Since $\gamma(\lambda^*, 0) = 0$ and using (3.5) we get that $L\Gamma x = 0$ for all $x \in V$, and hence $\Gamma x \in V \cap W$. Thus $\Gamma x = 0$ for all $x \in V$.

After these preliminaries, we can substitute

$$w = \gamma(\lambda, v) \tag{3.9}$$

in (3.3′) getting

$$P(F(\lambda, v + \gamma(\lambda, v))) = 0. \tag{3.10}$$

The equation (3.10) in the unknowns $(\lambda, v) \in \Lambda \times V$ is called *the bifurcation equation* and, together with (3.9), is equivalent (in $\Lambda \times V \times W$) to the initial equation $F(\lambda, u) = 0$.

Obviously, the preceding reduction is useful if the bifurcation equation is simpler than $F = 0$. This is the case when L is a Fredholm operator: if $\dim(V) = p$ and $\operatorname{codim}(R) = \dim(Z) = q$, then (3.10) is a system of q equations in the unknowns $(\lambda, v) \in \mathbb{R} \times \mathbb{R}^p$.

4 Bifurcation from the simple eigenvalue

In Section 1 we saw that the possible bifurcation points of $F(\lambda, u) = 0$ are those λ^* such that $F_u(\lambda^*, 0)$ is not invertible. To find sufficient conditions for λ^* to be a bifurcation point, some restrictions are in order. In this section we will study the case in which $L = F_u(\lambda^*, 0)$ is a Fredholm map with index zero and with one-dimensional kernel. In the case of equations like

$$G(u) = \lambda u,$$

with $G \in C^1(X, X)$, $G(0) = 0$ and $G'(0)$ compact, this corresponds to the case when λ^* is a *simple* eigenvalue of $G'(0)$.

Let us take a map $F \in C^2(\mathbb{R} \times X, Y)$ satisfying $F(\lambda, 0) = 0$ for all λ. We note explicitly that the condition $F \in C^2$ could be weakened assuming that $F \in C^1(\mathbb{R} \times X, Y)$ and has mixed partial derivative $F_{u,\lambda}$, see Remark 4.3 (i).

The hypothesis that $L = F_u(\lambda^*, 0)$ satisfies assumptions (a) and (b) of Section 3 needs to be specified here.

Keeping the notation of the preceding section, we set $V = \operatorname{Ker}(L)$, $R = R(L)$ and let W and Z denote complementary subspaces of V in X and R in Y, respectively.

We will say that L (or F) satisfies Assumption (I) if

(I-i) V is one-dimensional: $\exists u^* \in X, u^* \neq 0$ such that $V = \{tu^* : t \in$
        $\mathbb{R}\}$,
(I-ii) R is closed and codim $(R) = 1$.
According to (I-ii) Z is one-dimensional and there exists a linear func-
tional $\psi \in Y^*, \psi \neq 0$, such that

$$R = \{y \in Y : \langle \psi, y \rangle = 0\}.$$

We also use symbols P and Q to denote the projections onto Z and R,
respectively.

With this notation, the bifurcation equation (see (3.10)) becomes

$$\langle \psi, F(\lambda, tu^* + \gamma(\lambda, tu^*)) \rangle = 0. \tag{4.1}$$

It is convenient to set $\lambda = \lambda^* + \mu$, and

$$\beta(\mu, t) = \langle \psi, F(\lambda^* + \mu, tu^* + \gamma(\lambda^* + \mu, tu^*)) \rangle.$$

Note that β is a real-valued function defined in a neighbourhood U of
$(0, 0) \in \mathbb{R} \times \mathbb{R}$ and is of class C^2 there, because F and γ are C^2.

The following properties of β will be used later (subscripts denote
partial derivatives):
(β1) $\beta(\mu, 0) = 0$ for all μ; in particular,
(β2) $\beta_\mu(0, 0) = \beta_{\mu,\mu}(0, 0) = 0$;
(β3) $\beta_t(0, 0) = 0$.
To prove (β1) we note that

$$\beta(\mu, 0) = \langle \psi, F(\lambda^*, \gamma(\lambda^* + \mu, 0)) \rangle.$$

Since $\gamma(\lambda, 0) \equiv 0$ (see (3.7)) and $F(\lambda^*, 0) = 0$, (β1) and (β2) follow.

Next, to prove (β3) we differentiate β with respect to t yielding

$$\beta_t(\mu, t) = \langle \psi, F_u(\lambda^* + \mu, tu^* + \gamma(\lambda^* + \mu, tu^*))[u^* + \gamma_v(\lambda^* + \mu, tu^*)u^*] \rangle.$$

Letting $t = 0$ and taking into account that $\gamma(\lambda^* + \mu, 0) = 0$, we find

$$\beta_t(\mu, 0) = \langle \psi, F_u(\lambda^* + \mu, \gamma(\lambda^* + \mu, 0))[u^* + \gamma_v(\lambda^* + \mu, 0)u^*] \rangle$$

$$= \langle \psi, F_u(\lambda^* + \mu, 0)[u^* + \gamma_v(\lambda^* + \mu, 0)u^*] \rangle. \tag{4.2}$$

Since $\gamma_v(\lambda^*, 0) = 0$ (see (3.8)), we infer

$$\beta_t(0, 0) = \langle \psi, F_u(\lambda^*, 0)u^* \rangle = \langle \psi, Lu^* \rangle = 0,$$

proving (β3).

Furthermore, from (4.2) it follows that

$$\beta_{t,\mu}(\mu, 0) = \langle \psi, F_{u,\lambda}(\lambda^* + \mu, 0)[u^* + \gamma_v(\lambda^* + \mu, 0)u^*] \rangle$$

$$+ \langle \psi, F_u(\lambda^* + \mu, 0)\gamma_{v,\lambda}(\lambda^* + \mu, 0)[u^*] \rangle.$$

Here and always hereafter, we identify, according to Remark 1.4.4, the
mixed derivative such as $F_{u,\lambda}$ or $\gamma_{v,\lambda}$ with linear maps.

Letting $\mu = 0$ one finds

$$\beta_{t,\mu}(0,0) = \langle \psi, F_{u,\lambda}(\lambda^*,0)[u^* + \gamma_v(\lambda^*,0)u^*] \rangle$$

$$+ \langle \psi, F_u(\lambda^*,0)\gamma_{v,\lambda}(\lambda^*,0)[u^*] \rangle$$

$$= \langle \psi, F_{u,\lambda}(\lambda^*,0)[u^*] \rangle + \langle \psi, F_u(\lambda^*,0)\gamma_{v,\lambda}(\lambda^*,0)[u^*] \rangle.$$

Finally, since $\psi|_R = 0$ and $F_u(\lambda^*,0)\gamma_{v,\lambda}(\lambda^*,0)[u^*] \in R$,

$$\langle \psi, F_u(\lambda^*,0)\gamma_{v,\lambda}(\lambda^*,0)[u^*] \rangle = 0,$$

and we infer

(β4) $\beta_{t,\mu}(0,0) = \langle \psi, F_{u,\lambda}(\lambda^*,0)[u^*] \rangle.$

Furthermore, let us remark for future reference that, with direct calculations, one finds

(β5) $\beta_{t,t}(0,0) = \langle \psi, F_{u,u}(\lambda^*,0)[u^*, u^*] \rangle.$

We are now in position to state the main result of this section.

Theorem 4.1   *Suppose $F \in C^2(\mathbb{R} \times X, Y)$ be such that $F(\lambda,0) = 0$ for all $\lambda \in \mathbb{R}$. Let λ^* be such that $L = F_u(\lambda^*,0)$ satisfies assumption (I). Moreover, letting M denote the linear map $F_{u,\lambda}(\lambda^*,0)$, we assume that*

$$Mu^* \notin R. \tag{4.3}$$

Then λ^* is a bifurcation point for F. In addition the set of non-trivial solutions of $F = 0$ is, near $(\lambda^*,0)$, a unique C^1 cartesian curve with parametric representation on V.

Proof. According to the preceding discussion we have to solve the equation

$$\beta(\mu,t) = 0,$$

where β is C^2. In order to use the elementary Implicit Function Theorem, we need to "desingularize" β. For this, let us introduce the function

$$h(\mu,t) = \begin{cases} \beta(\mu,t)/t & \text{for } t \neq 0, \\ \beta_t(\mu,0) & \text{for } t = 0. \end{cases}$$

Using properties (β1–4) it is easy to see that h is $C^1, h(0,0) = 0$ and that

$$h_\mu(0,0) = \beta_{t,\mu}(0,0),$$

$$h_t(0,0) = \frac{1}{2}\beta_{t,t}(0,0).$$

Setting

$$a := h_\mu(0,0) \text{ and } b := h_t(0,0)$$

and using (β4) and (β5) we find

$$a = \langle \psi, Mu^* \rangle,$$

$$b = \frac{1}{2} \langle \psi, F_{u,u}(\lambda^*, 0)[u^*, u^*] \rangle.$$

In particular, from Assumption (4.3) one deduces

$$a = \langle \psi, Mu^* \rangle \neq 0. \tag{4.4}$$

Therefore the Implicit Function Theorem applies to $h = 0$ yielding a neighbourhood $(-\varepsilon, \varepsilon)$ of $t = 0$ and a unique function $\mu \in C^1(-\varepsilon, \varepsilon)$ such that $\mu(0) = 0$ and $h(\mu(t), t) = 0$ for all $t \in (-\varepsilon, \varepsilon)$. Since the equation $h(\mu, t) = 0$ is equivalent for $t \neq 0$ to $\beta(\mu, t) = 0$, it follows that the bifurcation equation (4.1) has been solved uniquely by $\mu = \mu(t)$.

Then, according to the results of Section 5.3, one finds that

$$F(\lambda^* + \mu(t), tu^* + \gamma(\lambda^* + \mu(t), tu^*)) = 0 \text{ for all } t \in (-\epsilon, \epsilon).$$

Note that $tu^* + \gamma(\lambda^* + \mu(t), tu^*) \neq 0$ provided $t \neq 0$. Therefore the set S of nontrivial solutions of $F(\lambda, u) = 0$ is given, in a neighbourhood of $(\lambda^*, 0)$, by the (unique) cartesian curve

$$\left.\begin{array}{l} \lambda = \lambda^* + \mu(t), \\ u = tu^* + \gamma(\lambda^* + \mu(t), tu^*), \end{array}\right\}$$

where $t \in (-\varepsilon, \varepsilon), t \neq 0$. This completes the proof of the theorem.

Theorem 4.1 becomes particularly expressive when $Y = X$ and

$$F(\lambda, y) = \lambda u - G(u),$$

where $G \in C^2(X, X)$ is such that $G(0) = 0$. As already seen in Section 1, the possible bifurcation points of f are points of the spectrum of $G'(0)$. Here we will show that, when $G'(0)$ is compact, any *simple* eigenvalue $\lambda \neq 0$ of $G'(0)$ is in fact a bifurcation point. A statement of this sort will provide a first answer to the question posed in Section 1, after Proposition 1.3.

Theorem 4.2 *Let $G \in C^2(X, X)$ be such that $G(0) = 0$ and such that $G'(0)$ is compact. Suppose that $\lambda^* \neq 0$ is a simple eigenvalue of $G'(0)$, in the sense that*

$$\dim(\mathrm{Ker}(\lambda^* I - G'(0))) = 1, \tag{4.5}$$

$$\mathrm{Ker}(\lambda^* I - G'(0)) \cap R(\lambda^* I - G'(0)) = \{0\}. \tag{4.6}$$

Then λ^* is a bifurcation point for $F(\lambda, u) = \lambda u - G(u)$.

Proof. Here $L = F_u(\lambda^*, 0) = \lambda^* I - G'(0)$, $V = \mathrm{Ker}(\lambda^* I - G'(0)) = \{tu^* :$

$t \in \mathbb{R}\}$ and $R = R(\lambda^* I - G'(0))$. Since $G'(0)$ is compact, assumption (I) follows immediately from (4.5).

Moreover, owing to the specific form of F, one has $M = F_{u,\lambda}(\lambda^*, 0)$ is the identity map, and therefore $a = \langle \psi, u^* \rangle$ where, as before, u^* denotes a vector spanning V. According to (4.6), $u^* \notin R$ and thus $a \neq 0$, proving (4.3). Then Theorem 4.1 applies and the result follows.

Remarks 4.3

(i) As anticipated, F can be assumed of class C^1, with continuous mixed partial derivative $F_{u,\lambda}$. For a proof, which requires some technicality, see [**PA**]. Similarly, one can show that Theorem 4.2 holds provided $G \in C^1(X, X)$.

(ii) The bifurcation results stated in Theorems 4.1 (or 4.2) do not hold, in general, if we only assume that L satisfies (I), namely that the dimension of $\mathrm{Ker}(L)$ and the codimension of $\mathrm{R}(L)$ are 1 without assuming (4.3) or (4.6). To see this, we can slightly modify Example 1.6. To be precise, let $X = Y = \mathbb{R}^2$ and

$$F(\lambda, x, y) = \begin{pmatrix} \lambda x - y - y^3 \\ \lambda y + x^3 \end{pmatrix}.$$

The same calculations performed in Example 1.6 show that, if

$$F(\lambda, x, y) = 0,$$

then there results

$$y^2 + y^4 + x^4 = 0.$$

Hence $F(\lambda, x, y) = 0$ has the trivial solution, only, and there are no bifurcation points.

Here, the derivative of F with respect to $u = \begin{pmatrix} x \\ y \end{pmatrix}$ evaluated at $(\lambda, 0, 0)$ is the map

$$F_u(\lambda, 0, 0) : \begin{pmatrix} x \\ y \end{pmatrix} \to \begin{pmatrix} \lambda x - y \\ \lambda y \end{pmatrix}.$$

Then, for $\lambda^* = 0$, we have that L can be identified with the matrix

$$L = \begin{bmatrix} 0 & -1 \\ 0 & 0 \end{bmatrix}.$$

Thus $V = R = \mathrm{span}\{u^*\}$, with $u^* = \begin{pmatrix} 1 \\ 0 \end{pmatrix}$, and (I) holds. On the other hand, M is the identity and hence $Mu^* = u^* \in R$. Note that Theorem 4.2 does not apply either. Indeed, the algebraic multiplicity of $\lambda = 0$ is 2, not 1 (in other words, (4.6) is not satisfied).

It is worth noticing that Theorem 4.2 applies only to maps of the specific form $F(\lambda, u) = \lambda u - G(u)$. We mean that when $X = Y$ but F is not in the form $\lambda I - G$, one has to use Theorem 4.1. In such a case, the condition $V \cap R = \{0\}$ does not play any role. To explain this, let us consider the map $F : \mathbb{R} \times \mathbb{R}^2 \to \mathbb{R}^2$,

$$F(\lambda, x, y) = \begin{pmatrix} \lambda x - y - y^3 \\ \lambda x + \lambda y + x^3 \end{pmatrix}.$$

As before, for $\lambda = 0$, $L = \begin{bmatrix} 0 & -1 \\ 0 & 0 \end{bmatrix}$ and $V = R = \text{span}\begin{pmatrix} 1 \\ 0 \end{pmatrix}$, but now the mixed derivative $F_{u,\lambda}(0,0,0)$ is the matrix $M = \begin{bmatrix} 1 & 0 \\ 1 & 1 \end{bmatrix}$ and $M : \begin{pmatrix} 1 \\ 0 \end{pmatrix} \to \begin{pmatrix} 1 \\ 1 \end{pmatrix} \notin R$, so that (4.3) holds true and $\lambda^* = 0$ is a bifurcation point for F. With a rather elementary calculation one can solve the system

$$\left. \begin{array}{l} \lambda x - y - y^3 = 0, \\ \lambda(x + y) + x^3 = 0, \end{array} \right\}$$

showing that the bifurcating branch has equation

$$\left. \begin{array}{l} y = -x^3 + \dots, \\ \lambda = -x^2 - x^4 + \dots. \end{array} \right\}$$

(iii) When $a = \langle \psi, F_{u,\lambda}(\lambda^*, 0)[u^*] \rangle = 0$ several different situations can occur and a more careful analysis is required. In the analytic case, it can be useful to employ the Newton polygon method to solve the bifurcation equation. For more details on this matter we refer to [**VT**]. The Newton polygon method has been extended to differentiable functions by Dieudonné [**D2**].

(iv) Assuming F more regular (say C^∞, for simplicity) one can complete Theorems 4.1 and 4.2 by some calculations which will allow us to specify the behaviour of the bifurcating branch near $(\lambda^*, 0)$ (see Figure 5.8).

Since $\mu(t)$ solves $h(\mu, t) = 0$, there results (see the proof of Theorem 4.1)

$$\mu'(0) = -\frac{h_t(0,0)}{h_\mu(0,0)} = -\frac{b}{a},$$

where (see earlier)

$$a = \langle \psi, F_{u,\lambda}(\lambda^*, 0)[u^*] \rangle,$$

$$b = \frac{1}{2} \langle \psi, F_{u,u}(\lambda^*, 0)[u^*, u^*] \rangle.$$

Therefore if $b \neq 0$ we have

$$\lambda = \lambda^* - \frac{b}{a} t + o(t)$$

and the bifurcating branch $(\lambda, u) \in \mathcal{S}$ can be parametrized (for $|\lambda - \lambda^*|$ small) in the form

$$u = -\frac{a}{b}(\lambda - \lambda^*)u^* + o(\lambda - \lambda^*).$$

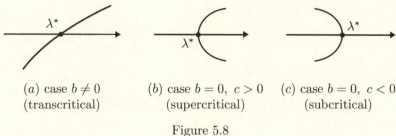

(a) case $b \neq 0$ (b) case $b = 0$, $c > 0$ (c) case $b = 0$, $c < 0$
(transcritical)   (supercritical)   (subcritical)

Figure 5.8

We note that when $b \neq 0$ the equation $F = 0$ has nontrivial solutions both for $\lambda > \lambda^*$ and for $\lambda < \lambda^*$ (*transcritical* bifurcation).

When $b = 0$ one finds

$$2c := \mu''(0) = -\frac{1}{3a}\langle \psi, F_{uuu}(\lambda^*, 0)[u^*]^3 \rangle. \qquad (4.7)$$

If $b = 0$ and $c \neq 0$ the bifurcating branch has the form

$$u = \pm \left(\frac{\lambda - \lambda^*}{c}\right)^{1/2} \cdot u^* + O(\lambda - \lambda^*).$$

Note that the preceding formula shows that if $c > 0$ (respectively, $c < 0$) then the bifurcating branch emanates on the right (respectively, left) of λ^* (*supercritical*, respectively *subcritical*, bifurcation).

It is worth remarking that, when $F(\lambda, u) = \lambda u - G(u)$ and G is smooth, the values of b and c are given by the formulas

$$b = -\frac{1}{2a}\langle \psi, G''(0)[u^*, u^*] \rangle$$

and

$$c = \frac{1}{6a}\langle \psi, G'''(0)[u^*]^3 \rangle.$$

Remark 4.4 When $F(\lambda, u) = \lambda u - G(u)$, with G compact, that is, $G(u_n)$ is relatively compact in X for any bounded sequence $\{u_n\}$, it is possible to use the Leray–Schauder topological degree and Theorem 4.2 can be greatly improved. Results of this sort are outside the scope of this book and cannot be discussed here. However, owing to their relevance, we shall give a review of the most important ones in a short appendix at the end of this chapter.

Postponing further examples to the next chapter we discuss here some problems related to those studied in Section 2.

Example 4.5 (Sturm–Liouville problems) Let $J = [0, \pi]$, $\alpha \in C^1(J)$, $\beta \in C(J)$, $\alpha, \beta > 0$ on J, $p \in C^2(J \times \mathbb{R} \times \mathbb{R})$ and let a_0, b_0, a_1, b_1 be such that $(a_0^2 + b_0^2)(a_1^2 + b_1^2) \neq 0$.

Consider the Sturm–Liouville b.v.p.

$$-\mathcal{L}u := -\frac{\mathrm{d}}{\mathrm{d}x}\left(\alpha\frac{\mathrm{d}}{\mathrm{d}x}u\right) + \beta u = \lambda u + p\left(x, u, \frac{\mathrm{d}u}{\mathrm{d}x}\right), x \in J, \qquad (4.8)$$

$$a_0 u(0) + b_0 u'(0) = a_1 u(\pi) + b_1 u'(\pi) = 0, \qquad (4.8')$$

where λ is a real parameter.

Setting $X = \{u \in C^2(J) : u \text{ satisfies } (4.8')\}, Y = C(J)$, define $F :$ $\mathbb{R} \times X \to Y$ by

$$F(\lambda, u) = \mathcal{L}u + \lambda u + p(u) \qquad (4.9)$$

(as usual we are using the same symbol p to indicate the Nemitski operator associated with the real-valued function p) so that the solutions of (4.8)–(4.8') are the pairs $(\lambda, u) \in \mathbb{R} \times X$ such that $F(\lambda, u) = 0$.

Suppose that $p = p(x, s, \xi)$ satisfies

$$p(x, 0, 0) \equiv 0, p_s(x, 0, 0) \equiv 0 \text{ and } p_\xi(x, 0, 0) \equiv 0.$$

As a consequence, one has

$$F(\lambda, 0) = 0 \text{ for all } \lambda,$$

$$F_u(\lambda, 0) : u \to \mathcal{L}u + \lambda u.$$

Recall (see subsection 0.6) that the linear problem

$$\left. \begin{array}{l} -\mathcal{L}u(x) = \lambda u(x) \ (x \in J), \\ a_0 u(0) + b_0 u'(0) = a_1 u(\pi) + b_1 u'(\pi) = 0, \end{array} \right\} \qquad (4.10)$$

has a sequence λ_k of positive, *simple* eigenvalues, such that $\lambda_k \to \infty$ as $k \to \infty$. Let φ_k be an eigenfunction of (4.10) corresponding to λ_k, normalized by

$$\int_0^\pi \varphi_k^2 \mathrm{d}x = 1.$$

Let us apply Theorem 4.1 with $\lambda^* = \lambda_k$ and $u^* = \varphi_k$. According to subsection 0.4, one has

$$V = \mathrm{Ker}[F_u(\lambda_k, 0)] = \mathbb{R}\varphi_k, \ R = R[F_u(\lambda_k, 0)] = \{u \in Y : \int_0^\pi u\varphi_k \mathrm{d}x = 0\}.$$

Therefore (I) holds true. Furthermore we can define ψ by $\langle \psi, u \rangle = \int_0^\pi u\varphi_k \mathrm{d}x$.

Since $F_{u,\lambda}(\lambda_k, 0) : v \to v$, $a = \langle \psi, \varphi_k \rangle = \int_0^\pi \varphi_k^2 \mathrm{d}x = 1$, proving (4.3). In conclusion, applying Theorem 4.1 we infer that each λ_k is a bifurcation point for $F = (4.9)$. Hence

For each $k = 1, 2, \ldots$, there is a continuous family u_λ of nontrivial solutions of (4.8)–(4.8') such that $\|u_\lambda\|_{C^2} \to 0$ as $\lambda \to \lambda_k$.

Example 4.6 (Dirichlet Problems) Let Ω be an open bounded domain in $\mathbb{R}^n$ and consider the boundary-value problem

$$\left.\begin{aligned} -\Delta u = \lambda u + p(x, u, \nabla u) \text{ in } \Omega, \\ u = 0 \text{ on } \partial\Omega, \end{aligned}\right\}$$

where $p \in C^2(\mathbb{R} \times \mathbb{R} \times \mathbb{R}^n)$ satisfies $p(x, 0, 0) = 0$, $p_s(x, 0, 0) = 0$ and $p_\xi(x, 0, 0) = 0$.

Since the discussion does not differ from that of the Sturm–Liouville problem, we will be sketchy.

Let $X = \{u \in C^{2,\alpha}(\overline{\Omega}) : u = 0 \text{ on } \partial\Omega\}$, $Y = C^{0,\alpha}(\overline{\Omega})$ and $F(\lambda, u) = \Delta u + \lambda u + p(u)$; one has that $F(\lambda, 0) = 0$ for all λ and $F_u(\lambda, 0)$ is the map $v \to \Delta v + \lambda v$. Hence $F_u(\lambda, 0)$ has a nontrivial kernel provided λ is an eigenvalue of

$$\left.\begin{aligned} -\Delta v = \lambda v \text{ in } \Omega, \\ v = 0 \text{ on } \partial\Omega. \end{aligned}\right\} \tag{4.11}$$

If λ_k is any *simple* eigenvalue of (4.11) with corresponding eigenfunction φ_k, normalized by $\int_\Omega \varphi_k^2 dx = 1$, then (I) holds true. As before, one has $R = R(F_u(\lambda_k, 0)) = \{u \in Y : \langle \psi, u \rangle = \int_\Omega u\varphi_k dx = 0\}$; since $F_{u,\lambda}(\lambda_k, 0) : v \to v, a = \langle \psi, \varphi_k \rangle = 1$ and (4.3) holds, too. Therefore, from Theorem 4.1 it follows that

> *any simple eigenvalue of* (4.11) *is a bifurcation point for* $F(\lambda, u) = \Delta u + \lambda u + p(u)$.

We note that, in particular, this result applies when we take the first eigenvalue of (4.11).

To know the behaviour of the bifurcating branch we refer to Remark 4.5. For simplicity, let us take a (smooth) nonlinearity p depending on u only.

Since here $F_{u,u}(\lambda_k, 0) : (v, w) \to p''(0)vw$, (4.4) becomes

$$b = \frac{1}{2}\langle \psi, F_{u,u}(\lambda^*, 0)[\varphi_k, \varphi_k] \rangle = \frac{1}{2}p''(0)\int_\Omega \varphi_k^3 dx.$$

If $p''(0) = 0$, one uses (4.7) yielding

$$c = -\frac{1}{3}p'''(0)\int_\Omega \varphi_k^4 dx.$$

For example, if $p(u) = -u^3$, then $b = 0$ and $c > 0$; hence the bifurcation is *supercritical* that is, occurs for $\lambda > \lambda_k$, while, if $p(u) = u^3$, then $c < 0$ and the bifurcation is *subcritical* (see Figure 5.9).

We end this section with some further remarks on the geometric character of Theorems 4.1–4.2.

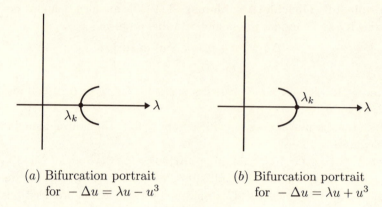

(*a*) Bifurcation portrait   (*b*) Bifurcation portrait
   for $-\Delta u = \lambda u - u^3$ for $-\Delta u = \lambda u + u^3$

Figure 5.9

After the Lyapunov–Schmidt reduction, the problem of finding the bifurcation points of $F(\lambda, u) = 0$ is reduced to the search for the zeros of a real-valued C^2 function $\beta = \beta(\mu, t)$, with the properties that

$$\beta(\mu, 0) = 0 \text{ for all } \mu,$$

$$\beta_t(0, 0) = 0,$$

$$\beta_{\mu, t}(0, 0) \neq 0.$$

The proof we have carried out led us to find two branches of solutions: that of the trivial zeros, and that of the nontrivial solutions, giving rise to the bifurcation branch. Suppose now that F is perturbed through $\tilde{F}$, with $\|F - \tilde{F}\|_{C^2} < \varepsilon$, with ε small. Perturbing F through $\tilde{F}$ will affect the bifurcation equation in the sense that $\beta = 0$ will be replaced by a perturbed bifurcation equation $\tilde{\beta} = 0$, with $\|\beta - \tilde{\beta}\|_{C^2}$ small. In general it is possible to prove that the zeros of $\tilde{\beta}$ become two branches that do not cross themselves, in general, but are merely "close" (see Figure 5.10).

This kind of perturbation phenomenon arises, for example, when one deals with bifurcation problems from the point of view of *numerical analysis*: approximation or truncation procedures can be viewed as perturbation.

For a discussion of this kind of problems, we refer to the paper by Golubitsky and Schaeffer [**GS**].

5 A bifurcation theorem from a multiple eigenvalue

In this section we will discuss a result dealing with a case in which $\mathrm{Ker}(L)$ is, possibly, not one-dimensional.

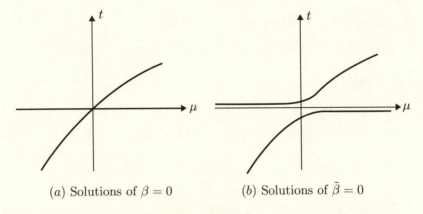

(a) Solutions of $\beta = 0$ (b) Solutions of $\tilde{\beta} = 0$

Figure 5.10

For simplicity, we consider an $F \in C^{\infty}(\mathbb{R} \times X, Y)$ and assume that (a) and (b) of Section 3 hold true. Keeping the notation of §3, we set $L = F_u(\lambda^*, 0)$ and write $X = V \oplus W, Y = Z \oplus R$, with $V = \operatorname{Ker}(L)$ and $R = R(L)$.

Let M denote the linear map $F_{u,\lambda}(\lambda^*, 0)$ (recall Remark 1.4.4) and $\mathcal{B}$ the bilinear map $F_{u,u}(\lambda^*, 0)$; then, setting $\lambda = \lambda^* + \mu$ we find that the equation $F = 0$ becomes

$$Lu + \mu M u + \frac{1}{2}\mathcal{B}(u, u) + \psi(\lambda^* + \mu, u) = 0, \tag{5.1}$$

where ψ is smooth and such that

$$\psi(\lambda, 0) \equiv 0, \ \psi_u(\lambda^*, 0) = 0, \ \psi_{u,u}(\lambda^*, 0) = 0, \ \psi_{\lambda,u}(\lambda^*, 0) = 0. \tag{5.2}$$

We seek solutions of the form $u = \mu(v + w)$, with $v \in V$ and $w \in W$.

Substituting into (5.1) we find

$$\mu L w + \mu^2 M(v + w) + \frac{1}{2}\mu^2 \mathcal{B}[v + w, v + w] + \psi(\lambda^* + \mu, \mu(v + w)) = 0.$$

Then, according to the Lyapunov–Schmidt reduction, the equation $F = 0$ is equivalent to the system

$$\mu^2 PM(v + w) + \frac{1}{2}\mu^2 P\mathcal{B}[v + w, v + w] + P\psi(\lambda^* + \mu, \mu(v + w)) = 0, \tag{5.3$'$}$$

$$\mu L w + \mu^2 QM(v + w) + \frac{1}{2}\mu^2 Q\mathcal{B}[v + w, v + w] + Q\psi(\lambda^* + \mu, \mu(v + w)) = 0, \tag{5.3$''$}$$

where P and Q indicate, as usual, the projections onto Z and R.

According to (5.2) we can write

$$\psi(\lambda^* + \mu, \mu(v + w)) = \mu^3 \tilde{\psi}(\mu, v, w)$$

where $\tilde{\psi}$ is smooth.

Hence (5.3')–(5.3'') are equivalent for $\mu \neq 0$ to

$$PM(v+w) + \frac{1}{2}PB[v+w, v+w] + \mu P\tilde{\psi}(\lambda^* + \mu, \mu(v+w)) = 0, \quad (5.4')$$

$$Lw + \mu QM(v+w) + \frac{1}{2}\mu QB[v+w, v+w] + \mu^2 Q\tilde{\psi}(\lambda^* + \mu, \mu(v+w)) = 0. \quad (5.4'')$$

With $\tilde{\Phi} = \tilde{\Phi}(\mu, v, w)$ denoting the left-hand side of (5.4'') there results $\tilde{\Phi}(0, v, 0) = 0$ for all $v \in V$ as well as $\tilde{\Phi}_w(0, v, 0) = L|_W$; hence, for any fixed $v^* \in V$, we can solve (5.4'') uniquely with respect to w in a neighbourhood of $\mu = 0, v = v^*$; from (5.4'') it follows readily that

$$w = \mu\gamma(\mu, v)$$

with γ smooth. Substituting into (5.4') we find the bifurcation equation

$$N(\mu, v) := PM(v + \mu\gamma(\mu, v)) + \frac{1}{2}PB[v + \mu\gamma(\mu, v), v + \mu\gamma(\mu, v)]$$

$$+ \mu P\tilde{\psi}(\lambda^* + \mu, \mu(v + \mu\gamma(\mu, v))) = 0. \quad (5.5)$$

Note that N is smooth. Moreover, let us point out that γ depends on $v^* \in V$. We will show that if v^* can be chosen in a suitable way then (5.5) can be solved, giving rise to a bifurcating branch for $F = 0$. More precisely one has the following.

Theorem 5.1 *Suppose that $V = \mathrm{Ker}(L)$ has a topological complement in X and $R = R(L)$ is closed and has a topological complement in Y. Moreover, letting $M = F_{u,\lambda}(\lambda^*, 0)$ and $B = F_{u,u}(\lambda^*, 0)$, suppose there exists $v^* \in V, v^* \neq 0$, such that*

(a) $PMv^* + \frac{1}{2}PB(v^*, v^*) = 0$,

(b)   *the linear map $S : V \to V$, $Sv = PMv + PB(v^*, v)$ is invertible. Then there is a branch of nontrivial solutions of $F = 0$ bifurcating from $(\lambda^*, 0)$ with equations*

$$\left.\begin{array}{c} \lambda = \lambda^* + \mu, \\[4pt] u = \chi(\mu), \end{array}\right\} \quad (5.6)$$

where $\chi(0) = 0$ and $\chi'(0) = v^*$.

Proof. From (a) it follows that $N(0, v^*) = 0$; moreover $N_v(0, v^*) = S$, which is invertible by (b). Then the Implicit Function Theorem applies to $N(\mu, v) = 0$. To be precise, there exists $v = v(\mu)$, defined for $|\mu|$ small, such that $v(0) = 0$ and

$$N(\mu, v(\mu)) = 0.$$

Hence we find a bifurcation branch of the form

$$u(\mu) = \mu(v(\mu) + \mu\gamma(\mu, v(\mu))).$$

Setting $\chi(\mu) := \mu(v(\mu) + \mu\gamma(\mu, v(\mu)))$, we get $\chi'(0) = v^*$. As a consequence, $u(\mu) \neq 0$, $|\mu|$ small and > 0, and $u = \chi(\mu)$ gives rise to a branch of nontrivial solutions of $F = 0$. This proves that λ^* is a bifurcation point for $F = 0$ and completes the proof of the theorem.

Remarks 5.2

(i) The branch found in the preceding theorem might not be unique: either because v^* might be not uniquely determined, or because there are other nontrivial solutions of $F = 0$, not in the form $\mu(v + w)$.

(ii) The equation of the bifurcating branch is parametrized with respect to μ and thus indicates that Theorem 5.1 gives rise to a transcritical bifurcation.

Theorem 5.1 can be used to find a sufficient condition for the existence of a bifurcation when $\dim(V) = \dim(Z) = 1$, but Theorem 4.1 does not apply.

Let $V = \mathbb{R}u^*$, and suppose that $PMu^* = 0$. Then conditions (a) and (b) of Theorem 5.1 become

(a') $PB[u^*, u^*] = 0$,

(b') the linear map $v \to PB[v^*, v]$ from V to Z is invertible.

If (a')–(b') hold true then an application of Theorem 5.1 yields the following.

Theorem 5.3 *Suppose $F \in C^2(\mathbb{R} \times X, Y)$ is such that $F(\lambda, 0) = 0$ for all $\lambda \in \mathbb{R}$. Let λ^* be such that $L = F_u(\lambda^*, 0)$ satisfies Assumption* (I) *and let $V = \mathbb{R}u*$. Moreover, we set $M = F_{u,\lambda}(\lambda, 0)$, $B = F_{u,u}(\lambda^*, 0)$ and we assume that $Mu^* \in \mathbb{R}$ and that* (a')–(b') *hold true. Then λ^* is a bifurcation point for F.*

An application

Let us apply Theorem 5.1 to the following problem: given a continuous 2π-periodic function h, to find 2π-periodic solutions of

$$u'' + \lambda u + hu^2 = 0, \tag{5.7}$$

We set $X = C_{2\pi}^2$, $Y = C_{2\pi}$, where $C_{2\pi}^k$ (resp. $C_{2\pi}$) denotes the space of 2π-periodic C^k functions (resp. continuous functions), and let $F : \mathbb{R} \times X \to Y$,

$$F(\lambda, u) = u'' + \lambda u + hu^2. \tag{5.8}$$

Here $Lv = v'' + \lambda v$, $Mv = v$ and $B[u, v] = 2huv$. For $\lambda = \lambda^* =$

k^2, $V = \mathrm{Ker}\,(L)$ is two-dimensional and spanned by $\{\cos kt, \sin kt\}$. Moreover, $Z = \mathrm{span}\,\{\cos kt, \sin kt\}$, too, and $R = R(L)$ is L^2-orthogonal to Z. As for the corresponding projection $P : Y \to Z$, one has that $Ph = (a_k \cos kt, b_k \sin kt)$ where

$$a_k = \frac{1}{\pi} \int_0^{2\pi} h(t) \cos kt \, dt,$$

$$b_k = \frac{1}{\pi} \int_0^{2\pi} h(t) \sin kt \, dt.$$

If we write $v = A \cos kt + B \sin kt$ condition (a) leads us to find nontrivial solutions of the system

$$A + \frac{1}{\pi} \int_0^{2\pi} (A \cos kt + B \sin kt)^2 h(t) \cos kt = 0,$$

$$B + \frac{1}{\pi} \int_0^{2\pi} (A \cos kt + B \sin kt)^2 h(t) \sin kt = 0.$$

This system is of the form

$$\left.\begin{aligned} A + \mathcal{P}(A, B) = 0, \\ B + \mathcal{Q}(A, B) = 0, \end{aligned}\right\} \tag{5.9}$$

where $\mathcal{P}, \mathcal{Q}$ are homogeneous polynomials of degree 2, whose coefficients depend on h. From the geometrical point of view, the solutions can be thought of as the intersections of two conics crossing through the origin transversally to each other. So they intersect in another point in the projective plane. This intersection is not on the "line at infinity", that is, $A, B \in \mathbb{R}$, provided the system

$$\left.\begin{aligned} \mathcal{P}(A, B) = 0, \\ \mathcal{Q}(A, B) = 0, \end{aligned}\right\}$$

has the trivial solution $A = B = 0$ only. It is easy to see that this is the case for all $h \in Y \backslash Y_0$, for some thin set Y_0 (in the sense of Baire). Then, for a "generic" h (5.9) has a nontrivial solution $(A^*, B^*) \in \mathbb{R}^2$; as for condition (b), it also holds for all h up to a thin set. Then we can conclude that, for all $h \in Y$, up to a set of first category in Y, each $\lambda = k^2, k = 1, 2, \ldots$, is a bifurcation for F given by (5.6); each bifurcating branch gives rise to a family of 2π-periodic solutions of (5.5).

Appendix

In this short appendix we want to review some very important bifurcation results, which require tools other than the Local Inversion Theorem.

We will deal with equations of the type

$$F(\lambda, u) = \lambda u - G(u) = 0, \tag{A1}$$

where G satisfies

(G1) $G \in C(X, X)$ and is differentiable at $u = 0$, with (compact) derivative $A = G'(0)$,
(G2) G is compact.
It is always understood that $G(0) = 0$.

Theorem A1 (Krasnoselskii, [Kr1]) *Suppose that* (G1–2) *hold and let* λ^* *be an eigenvalue of* A *with odd (algebraic) multiplicity. Then* λ^* *is a bifurcation point for* F.

Roughly, the proof relies on the following arguments. If, supposing the contrary, λ^* is not a bifurcation point then there exist a ball D around $u = 0$ and $\varepsilon > 0$ such that

$$F(\lambda, u) \neq 0 \text{ for all } \lambda \in [\lambda - \varepsilon, \lambda + \varepsilon], \text{ for all } u \in \partial D. \tag{A2}$$

In view of (A2), it makes sense to consider the Leray–Schauder topological degree, $\mathrm{d}(F_\lambda, D, 0)$, of $F_\lambda := F(\lambda, .)$, with respect to D and $u = 0$, and, by the homotopy invariance of the degree, one has

$$\mathrm{d}(F_{\lambda^* - \varepsilon}, D, 0) = \mathrm{d}(F_{\lambda^* + \varepsilon}, D, 0). \tag{A3}$$

On the other hand, if necessary taking D smaller, the degree of F_λ can be evaluated by linearization: more precisely, if λ is not an eigenvalue of $A = G'(0)$, then one has

$$\mathrm{d}(F_\lambda, D, 0) = \mathrm{d}(\lambda I - A, D, 0) = (-1)^k, \tag{A4}$$

where k denotes the sum of the (algebraic) multiplicities (see subsection 0.4) of the eigenvalues μ of A, with $\mu > \lambda$.

Let m denote the sum of the (algebraic) multiplicities of the eigenvalues μ of A, with $\mu > \lambda^*$, and m^* that of λ^*. If necessary taking ε smaller, we can assume that λ^* is the only eigenvalue of A in the interval $[\lambda - \varepsilon, \lambda + \varepsilon]$. Then from (A4) we infer

$$\mathrm{d}(F_{\lambda^* + \varepsilon}, D, 0) = (-1)^m,$$

$$\mathrm{d}(F_{\lambda^* - \varepsilon}, D, 0) = (-1)^{m + m^*}.$$

Since m^* is odd, it follows that $\mathrm{d}(F_{\lambda^* - \varepsilon}, D, 0) \neq \mathrm{d}(F_{\lambda^* + \varepsilon}, D, 0)$, in contradiction with (A3). This proves that λ^* is a bifurcation point.

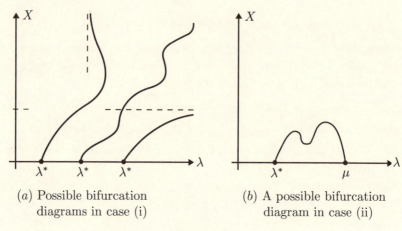

(*a*) Possible bifurcation   (*b*) A possible bifurcation
    diagrams in case (i)   diagram in case (ii)

Figure 5.11

Actually, the global nature of the topological degree can be used to improve Theorem A1 as follows.

Theorem A2 (Rabinowitz, [R2]) *Suppose that* (G1–2) *hold and let* λ^* *be an eigenvalue of A with odd (algebraic) multiplicity. Then from λ^* there branches off a continuum (namely a closed connected set) Σ of nontrivial solutions of $F = 0$ such that either*

(i) Σ *is unbounded, or*

(ii) Σ *meets another eigenvalue $\mu \neq \lambda^*$ of A.*

Theorem A2 applies to a large variety of problems. Among others, we mention Sturm–Liouville problems [**CrR**], existence of positive solutions of nonlinear eigenvalue problems [**AH**], existence of vortex rings in an ideal fluid [**AmiT**].

A last result which is worth recalling deals with the case in which G is a *variational operator*. To be precise, let us assume that X is a Hilbert space and that there exists $g : X \to \mathbb{R}$, such that $G = \nabla g$. Note that in such a case (A1) becomes $\nabla g(u) = \lambda u$, whose solutions can be found as *critical points* of g on the Hilbert sphere $\|u\| = \rho$, the parameter λ playing the role of the Lagrange multiplier.

Theorem A3 (Krasnoselski, [Kr1]) *Suppose $G \in C^1(X, X)$ is a variational operator and satisfies* (G2). *Then any eigenvalue of $A = G'(0)$ is a bifurcation point for $F = 0$.*

For a proof using Morse Theory, see [**MP**]. Improvements can be found in [**Bö**] and [**Mar**].

6

Bifurcation problems

There is a broad variety of problems arising in applications that can be handled by the Bifurcation Theorem stated in Section 5.4. In the present chapter we will discuss some of them. We have tried to choose problems that are relevant from the physical point of view but that do not need too much technicality. Only one of them, the Bénard Problem discussed in Section 2, is not self-contained. Indeed, the analysis of linearized equations requires some delicate tools that would need much more space. Nevertheless, the relevance of the problem has driven us to include it in this chapter, even if we had to be sketchy in several points.

1 The rotating heavy string

Following the formulation of Kolodner [Ko], we consider a string with uniform density ρ and length $= 1$, hung at the origin of the coordinates (the z-axis will be considered to be pointing downwords) in $\mathbb{R}^3$. The points on the string will be parametrized through the arclength $s \in [0, 1]$ and denoted by $\mathbf{x}(s, t) = (x(s, t), y(s, t), z(s, t))$. It is convenient to take s in such a way that $\mathbf{x}(1, t) = (0, 0, 0)$ is the fixed endpoint of the string.

The equations of the motion are

$$\rho \mathbf{x}_{tt} = \rho \mathbf{g} + (T \mathbf{x}_s)_s \tag{1.1}$$

together with

$$|\mathbf{x}_s|^2 = 1, \tag{1.2}$$

where $\mathbf{g} = (0, 0, g)$ is the acceleration of gravity, T denotes the tension and subscripts denote partial derivatives.

We look for solutions corresponding to a string moving with constant angular velocity $\omega > 0$. More precisely, we will seek for x, y, z and T of the form

$$x = x(s, t) = r(s) \cos \omega t,$$
$$y = y(s, t) = r(s) \sin \omega t,$$
$$z = z(s),$$
$$T = T(s).$$

In addition we will also require that $z'(s) < 0$.

There result

$$x_{tt} = -r\omega^2 \cos \omega t, \quad y_{tt} = -r\omega^2 \sin \omega t,$$
$$x_s = r' \cos \omega t, \quad y_s = r' \sin \omega t$$

where $\prime$ denotes d/ds. Substituting in (1.1)–(1.2) one finds easily

$$\left. \begin{array}{l} (Tr')' + \rho\omega^2 r = 0, \\ (Tz')' + \rho g = 0, \\ (r')^2 + (z')^2 = 1. \end{array} \right\} \tag{1.3}$$

The boundary conditions to be added to (1.3) are

$$\left. \begin{array}{l} r(1) = 0, \\ z(1) = 0, \\ T(0) = 0. \end{array} \right\} \tag{1.4}$$

It is convenient to modify (1.3) to obtain a single equation. First of all, the second of (1.3) is independent of r, and can be integrated; taking into account that $T(0) = 0$, one has

$$T(s)z'(s) = -\rho g s. \tag{1.5}$$

In particular, since $z'(s) < 0$, one has that $T(s) > 0$ for all $s > 0$.

From (1.5) and the last of (1.3) we infer

$$T^2 z'^2 = T^2(1 - r'^2) = (\rho g s)^2.$$

Setting

$$u(s) = \frac{T(s)r'(s)}{\rho g},$$

one finds

$$T^2 - (\rho g u)^2 = (\rho g s)^2, \tag{1.6}$$

That is, $T^2 = \rho^2 g^2(u^2 + s^2)$. Recalling that $T > 0$ for all $s > 0$, we find that (1.6) yields

$$T = \rho g \sqrt{(u^2 + s^2)}.$$

On the other hand, from the first of (1.3) and letting $\lambda = \omega^2/g > 0$ we deduce

$$u' + \lambda r = 0. \tag{1.7}$$

Differentiating, we find

$$0 = u'' + \lambda r' = u'' + \lambda \rho g \frac{u}{T},$$

and finally

$$u'' + \lambda \frac{u}{\sqrt{(u^2 + s^2)}} = 0.$$

As for the boundary conditions, one has $u(0) = T(0)r'(0)/\rho g = 0$, while from (1.7) and $r(1) = 0$ we infer $u'(1) = 0$.

In conclusion, the problem given in eqns (1.1), (1.2), (1.4), can be written in the form

$$\left.\begin{array}{l} u'' + \lambda \dfrac{u}{\sqrt{(u^2 + s^2)}} = 0 \text{ for } 0 < s < 1, \\[2mm] u(0) = u'(1) = 0. \end{array}\right\} \tag{1.8}$$

If (1.8) has a non-trivial solution $u(s)$, corresponding to a certain value $\lambda > 0$, then we can go back to the solutions of (1.1), (1.2), (1.4) by

$$r(s) = -\frac{u'(s)}{\lambda},$$

$$T(s) = \lambda g \sqrt{[u^2(s) + s^2]}, z(s) = \int_s^1 \frac{\tau d\tau}{\sqrt{[u^2(\tau) + \tau^2]}}, \omega = \sqrt{g\lambda}.$$

We remark that from the preceding formulas one finds

$$z'(s) = -s(u^2(s) + s^2)^{-1/2} < 0 \text{ for all } s < 0,$$

and hence the solution is consistent with our setting.

In order to employ the bifurcation results of Section 5.4, we fix the functional setting, taking $X = \{u \in C^2([0,1]) : u(0) = u'(1) = 0\}$ and $Y = C([0,1])$. It is convenient to introduce some further notation. For $u \in X$ we denote by $\tilde{u}(s)$ the function defined by

$$\tilde{u}(s) = \begin{cases} u(s)/s & \text{for } 0 < s \leq 1, \\ u'(0) & \text{for } s = 0, \end{cases}$$

in such a way that we can write

$$\frac{u(s)}{\sqrt{[u(s)^2 + s^2]}} = \frac{\tilde{u}(s)}{\sqrt{[\tilde{u}(s)^2 + 1]}}.$$

We also let $\Phi(u)$ denote the function

$$s \rightarrow \frac{\tilde{u}(s)}{\sqrt{[\tilde{u}(s)^2 + 1]}}, \tag{1.9}$$

and define $F : \mathbb{R}^+ \times X \to Y$ by setting

$$F(\lambda, u) = u'' + \lambda \Phi(u). \qquad (*)$$

Plainly, if $u \in X$, $\lambda > 0$ and $F(\lambda, u) = 0$ then u is a solution of (1.8).

Lemma 1.1 $\Phi \in C^\infty(X, Y)$ *and* $\Phi'(u)v$ *is the function*

$$s \to \frac{\tilde{v}}{(\tilde{u}^2 + 1)^{1/2}} - \frac{\tilde{u}^2 \tilde{v}}{(\tilde{u}^2 + 1)^{3/2}}.$$

In particular $\Phi'(0)v$ *is the function* $s \to \tilde{v}(s) = v(s)/s$.

Proof. It suffices to note that Φ is obtained by composition between the linear continuous map $u \to \tilde{u}$ (from X to Y) and the C^∞ map

$$w \to w/(w^2 + 1)^{1/2}$$

(from Y to itself).

From Lemma 1.1 we infer that the linearized problem $v'' - \lambda \Phi'(0)v = 0$ is given by

$$\left. \begin{aligned} v'' + \lambda \frac{v}{s} &= 0, \\ v(0) = v'(1) &= 0. \end{aligned} \right\} \qquad (1.10)$$

It is well known that the equation $v'' + \lambda v/s = 0$ is related with the Bessel equations.

In fact, setting

$$t = 2\sqrt{\lambda s}, \quad v(s) = t\,w(t), \qquad (1.11)$$

one readily has

$$v'' = \frac{4\lambda^2}{t} \left(\frac{d^2 w}{dt^2} + \frac{1}{t} \frac{dw}{dt} - \frac{w}{t^2} \right).$$

Then $v'' + \lambda v/s = 0$ becomes

$$t^2 \frac{d^2 w}{dt^2} + t \frac{dw}{dt} + (t^2 - 1)w = 0, \qquad (1.12)$$

which is the κth Bessel equation $t^2 z'' + t z' + (t^2 - \kappa^2)z = 0$, with $\kappa = 1$. In the following J_κ denotes the κth Bessel Function *of the first kind* (see [**Sa**] Cap.III, §6). For example, J_1 is defined by

$$J_1(t) = \frac{t}{2} \left(1 - \frac{1}{2!} \left(\frac{t}{2} \right)^2 + \frac{1}{2!3!} \left(\frac{t}{2} \right)^4 - \frac{1}{3!4!} \left(\frac{t}{2} \right)^6 + \cdots \right).$$

Recall that there results $J_\kappa'(t) = J_{\kappa-1}(t) - (\kappa/t)J_\kappa(t), \kappa \geq 1$ (see [**Sa**], p.175). In particular, one has

$$J_1'(t) = J_0(t) - \frac{1}{t} J_1(t). \qquad (1.13)$$

Lemma 1.2 *The eigenvalues of* (1.10) *are given by*
$$\lambda_n = (\sigma_n/2)^2,$$
where σ_n denotes the nth zero of the Bessel Function J_0. If $\lambda = \lambda_n$ then the corresponding eigenfunctions of (1.10) *are of the form $v(s) = c\varphi_n(s)$, $c \in \mathbb{R}$, where*
$$\varphi_n(s) = 2\sqrt{\lambda_n}s \cdot J_1(2\sqrt{\lambda_n}s).$$

Proof. Through the change of variable (1.11) the equation $v'' + \lambda v/s = 0$ translates into the Bessel equation (1.12). From the theory of the Bessel equations [**Sa**] it follows that the general solution $w(t)$ of (1.12) such that $tw(t) \to 0$ as $t \to 0$ is given by
$$w(t) = cJ_1(t), \; c \in \mathbb{R}.$$
Hence the general solution of
$$\left. \begin{array}{r} v'' + \lambda \dfrac{v}{s} = 0, \\[2mm] v(0) = 0, \end{array} \right\}$$
is given by
$$v(s) = 2c\sqrt{\lambda}s \cdot J_1(2\sqrt{\lambda}s).$$
From
$$v'(s) = c\sqrt{\frac{\lambda}{s}} \cdot J_1(2\sqrt{\lambda}s) + 2c\lambda J_1'(2\sqrt{\lambda}s)$$
and using (1.13) we get that
$$v'(s) = 2c\lambda J_0(2\sqrt{\lambda}s).$$
Then the boundary condition $v'(1) = 0$ gives rise to
$$2c\lambda J_0(2\sqrt{\lambda}) = 0.$$
If $2\sqrt{\lambda} \neq \sigma_n$ then $c = 0$ and $v \equiv 0$. Hence (1.10) has non-trivial solutions if and only if $\lambda_n = (\sigma_n/2)^2$ and the lemma follows.

We are now in position to state the following.

Theorem 1.3 *For $n = 1, 2, \ldots$, let σ_n denote the n-th zero of the Bessel function J_0 and let $\lambda_n = (\sigma_n/2)^2$. Then any λ_n is a bifurcation point for $F = 0$, where F is given by* (*).

Proof. We are going to apply Theorem 5.4.1 (jointly with Remark 5.4.3(i)) to $F(\lambda, u) = u'' + \lambda\Phi(u)$ (where $\Phi(u)$ is given by (1.9)) with $\lambda^* = \lambda_n$. According to Lemma 1.1, $F \in C^1(\mathbb{R} \times X, Y)$. Keeping the notation of the preceding chapter, we set $L = F_u(\lambda_n, 0)$, $V = \text{Ker}(L)$

and $R = R(L)$. By Lemma 1.2 it follows that V is one-dimensional and spanned by φ_n.

As for R, we can argue as follows. The equation $v'' + \lambda \tilde{v} = h$ ($h \in Y$) can be written in the form $v = \lambda G(\tilde{v}) - G(h)$, where G denotes the Green operator of $-\mathrm{d}^2/\mathrm{d}s^2$ with the boundary conditions $v(0) = v'(1) = 0$. Then the Fredholm Alternative Theorem 0.1 applies and R has codimension 1 in Y.

The preceding arguments show that Assumption (I) of Theorem 5.4.1 holds true.

Lastly, $a = \langle \psi, F_{u,\lambda}(\lambda_n, 0)[\varphi_n] \rangle = \int_0^1 \varphi_n^2 > 0$ and Theorem 5.4.1 applies yielding a branch of nontrivial solutions of (1.8) bifurcating from $(\lambda_n, 0)$.

Remarks 1.4

(i) As remarked before, nontrivial solutions of $F = 0$ correspond to nontrivial solutions of (1.8). Moreover, since $F(\lambda, u) = F(\lambda, -u)$, solutions arise in pairs $(u, -u)$. One could also readily show that the bifurcation is *supercritical*. Then, recalling also the discussion in Remark 5.4.3 (iv), we can say that *for $\lambda > \lambda_n$, λ near λ_n, there exists a family $(u_n(\lambda), -u_n(\lambda))$ of pairs of nontrivial solutions of (1.8), continuously depending on λ, such that $\|u_n(\lambda)\|_{C^2} \to 0$ as $\lambda \downarrow \lambda_n$.*

(ii) The preceding result can be improved: indeed, using shooting methods, it has been shown in [**Ko**] that *for any $\lambda_n < \lambda \le \lambda_{n+1}$ the problem (1.8) has exactly n pairs of nontrivial solutions $\pm u_1 \pm u_2, \dots, \pm u_n$ such that u_j has j isolated zeros in $[0,1]$.*

2 The Bénard problem

Fluid dynamics is a typical area of application for bifurcation theory. In this section we will discuss a very classical question: the Bénard Problem dealing with convective motions in a heated fluid. We shall restrict our treatment to the case of *planar stationary* flows; such a problem, even though much simpler than the general one, nevertheless is still physically interesting and quite nontrivial from the mathematical point of view.

For more complete results on the Bénard Problem, see, for example, [**R3**][**Ve**].

Consider a viscous fluid conducting heat, situated between two horizontal walls, kept at a constant temperature, and suppose the lower wall is warmer than the higher wall. If the difference in temperature is small, the fluid remains at rest and the transmission of heat occurs only by conduction; but if the temperature gradient passes a certain critical threshold, convective motions are triggered. This is indeed what was experimentally observed by Bénard around 1900.

The mathematical model we are going to consider (due to Boussinesq) is based on the following physical assumptions:

(a)   the fluid has constant density, constant heat conduction and satisfies the Navier–Stokes equations;

(b)   the only forces acting on the fluid are those of buoyancy (vertically directed and depending linearly on temperature).

Let us remark that the variations of density due to temperature are important for the generated buoyancy forces (condition (b)), but they are neglected in the dynamics of the fluid (condition (a)).

With $\mathbf{u}$ denoting the velocity of the fluid, p the pressure, θ the difference between the temperature and the linear function interpolating the values on the bounding walls, one finds the following system:

$$\left.\begin{aligned} &\mathbf{u}_t + (\mathbf{u} \cdot \nabla)\mathbf{u} + \nabla p - \Delta \mathbf{u} - R\theta \mathbf{e} = 0, \\ &\nabla \cdot \mathbf{u} = 0, \\ &\theta_t + P(\mathbf{u} \cdot \nabla)\theta - \Delta\theta - \mathbf{e} \cdot \mathbf{u} = 0. \end{aligned}\right\} \tag{2.1}$$

Here

   - $\mathbf{e}$ is a unit vector vertically directed,
   - R (Rayleigh number) is a positive constant, and is proportional to the gradient of the temperature,
   - P (Prandtl number) is another positive constant.

The form of system (2.1), containing two constants only, has been obtained by choosing suitable units. We can also suppose that the bounding walls are represented by planes at level -1 and $+1$, respectively. On these planes both the velocity field u and θ vanish.

For the sake of simplicity, we will be interested in *planar stationary* solutions only. Let us introduce cartesian coordinates (x, y) and set $\mathbf{u} = (v, w)$. Since $\mathbf{u}$ is a planar, solenoidal field, $\nabla \cdot \mathbf{u} = 0$, we can introduce a stream potential ψ by setting

$$v = \frac{\partial \psi}{\partial y}, \ w = -\frac{\partial \psi}{\partial x}.$$

Substituting in (2.1) one readily finds the system

$$\left.\begin{aligned} &\Delta^2\psi - R\frac{\partial\theta}{\partial x} + \mathcal{A}(\psi) = 0, \\ &-\Delta\theta + \frac{\partial\psi}{\partial x} + P\mathcal{B}(\psi, \theta) = 0, \end{aligned}\right\} \tag{2.2}$$

where $\mathcal{A}$ is a quadratic function in the derivatives of ψ up to the third order, and $\mathcal{B}$ is bilinear in the first derivatives of ψ and θ. The region where the system (2.2) is considered is the strip

$$S = \{(x, y) \in \mathbf{R}^2 : -\infty < x < \infty, -1 \le y \le 1\}.$$

The boundary conditions become

$$\left. \begin{aligned} \psi(x, 1) &= \psi(x, -1) = 0, \\ \psi_y(x, 1) &= \psi_y(x, -1) = 0, \\ \theta(x, 1) &= \theta(x, -1) = 0. \end{aligned} \right\} \tag{2.3}$$

We remark that the physical conditions of the problem demand that ψ be constant on ∂S, but we further require that the flow through any vertical section be zero: this implies that ψ has the *same* constant value on ∂S (obviously, such a value can be taken to be 0).

We look for nontrivial solutions of (2.2)–(2.3), which are $(2\pi/a)$-periodic in the x variable, where a is a positive, *fixed* value. The search for periodic solutions is suggested by the fact that, experimentally, convective motions have – in many cases – a roll pattern, repeated periodically.

Let us introduce a further restriction to the class of admissible functions; more precisely, let us assume that

ψ *is an even function of x and θ is an odd function of x.*
It is convenient to introduce the following function spaces:

H_a^4 is the space of functions ψ defined in S, $(2\pi/a)$-periodic in x, even in x, having fourth-order derivatives in $L^2(S)$ and such that $\psi(x, \pm 1) = \psi_y(x, \pm 1) = 0$,

H_a^2 is the space of functions θ defined in S, $(2\pi/a)$-periodic in x, odd in x, having second-order derivatives in $L^2(S)$ and such that $\theta(x, \pm 1) = 0$.

L_a' the space of functions defined in S, $(2\pi/a)$-periodic in x, even in x, and square-integrable,

L_a'' the space of functions defined in S, $(2\pi/a)$-periodic in x, odd in x, and square-integrable.

Let $X = H_a^4 \times H_a^2$ and $Y = L_a' \times L_a''$. An analysis of the terms contained in $\mathcal{A}(\psi)$ and $\mathcal{B}(\psi, \theta)$ shows that if $(\psi, \theta) \in H_a^4 \times H_a^2$ then $\mathcal{A}(\psi)$ and $\mathcal{B}(\psi, \theta)$ belong to H_a^1. If we take into account that the operator

$$A := \begin{bmatrix} \Delta^2 & 0 \\ 0 & -\Delta \end{bmatrix}$$

is elliptic, it follows by standard regularity theory that the solutions $(\psi, \theta) \in X$ of (2.2) are in fact smooth.

In addition, the same analysis shows that the map

$$(\psi, \theta) \to (\mathcal{A}(\psi), \mathcal{B}(\psi, \theta))$$

is of class C^∞ as a map from X to Y. Then the left-hand side of (2.2) defines, for all R, a map $F_R : H_a^4 \times H_a^2 \to L_a' \times L_a''$ which is of class C^∞. Taking the Rayleigh number R as bifurcation parameter, we are led to the equation $F(R, \psi, \theta) := F_R(\psi, \theta) = 0$. Plainly $F(R, 0, 0) = 0$ for all R.

In order to apply the bifurcation results of Section 5.4 we have to study the system obtained by linearization from (2.2)–(2.3), namely

$$\left. \begin{array}{l} \Delta^2 \psi - R \dfrac{\partial \theta}{\partial x} = 0, \\[2mm] -\Delta \theta + \dfrac{\partial \psi}{\partial x} = 0, \\[2mm] \psi(x, \pm 1) = \psi_y(x, \pm 1) = \theta(x, \pm 11) = 0. \end{array} \right\} \tag{2.4}$$

We want to show that (2.4) has a simple eigenvalue R such that Theorem 5.4.1 applies to F.

After the substitution $\psi = \sqrt{R} \cdot \phi$, (2.4) becomes

$$\left. \begin{array}{l} \Delta^2 \phi - \sqrt{R} \cdot \dfrac{\partial \theta}{\partial x} = 0, \\[2mm] -\Delta \theta + \sqrt{R} \cdot \dfrac{\partial \phi}{\partial x} = 0, \\[2mm] \psi(x, \pm 1) = \psi_y(x, \pm 1) = \theta(x, \pm 11) = 0. \end{array} \right\} \tag{2.5}$$

The left-hand side of (2.5) contains an operator L_R which can be written in matricial form as follows:

$$L = \begin{bmatrix} \Delta^2 & -\sqrt{R} \cdot \frac{\partial}{\partial x} \\ \sqrt{R} \cdot \frac{\partial}{\partial x} & -\Delta \end{bmatrix}$$

$$= \begin{bmatrix} \Delta^2 & 0 \\ 0 & -\Delta \end{bmatrix} + \sqrt{R} \cdot \begin{bmatrix} 0 & \frac{\partial}{\partial x} \\ \frac{\partial}{\partial x} & 0 \end{bmatrix} = A + \sqrt{R} \cdot B$$

The operator A, as an operator on the Hilbert space $L_a' \times L''$ with domain of definition $H_a^4 \times H_a^2$, is self-adjoint and invertible. It is easy to verify that the operator B is symmetric; since, for real λ big enough, the operator $L_R + \lambda I$ (still with domain $H_a^4 \times H_a^2$) is invertible, we deduce that L_R is self adjoint [†]. Therefore, Range(L) is the subspace orthogonal to Ker(L).

[†] Here we use the following abstract result. Let H be a Hilbert space and let A, B two linear maps defined on a dense domain $\mathcal{D} \subset H$. Suppose that (i) A is a self-adjoint; (ii) B is continuous and symmetric, in the sense that $(Bu|v) = (u|Bv)$ for all $u, v \in \mathcal{D}$, and (iii) $T = \lambda I + A + B$ is invertible from $\mathcal{D}$ to H for some real λ. Then $A + B$ is self-adjoint.

In order to verify the assumptions of Theorems 5.4.1 it remains to prove that, for suitable values of R, $\mathrm{Ker}(L_R)$ is one-dimensional.

In view of the properties of the spaces H_a^4, H_a^2, we can set

$$\phi(x,y) = \phi_o(y) + \Sigma_{k \geq 1} \cos(kax)\,\phi_k(y),$$
$$\theta(x,y) = \Sigma_{k \geq 1} \sin(kax)\theta_k(y).$$

Substituting into (2.5), one readily obtains the following system of infinitely many equations:

$$\left.\begin{array}{l} M_{ka}^2 \phi_k - \sqrt{R} \cdot ka \; \theta_k = 0, \\[4pt] M_{ka}\theta_k - \sqrt{R} \cdot ka \; \phi_k = 0, \\[4pt] \phi_k(\pm 1) = \phi'_k(\pm 1) = \theta_k(\pm 1) = 0, \end{array}\right\} \qquad (2.6)_k$$

where $M_{ka} = -\mathrm{d}^2/\mathrm{d}y^2 + k^2 a^2$, and $k = 1, 2, \ldots$.

Let us consider, in general, the system

$$\left.\begin{array}{l} M_{ka}^2 \phi - \lambda\theta = 0, \\[4pt] M_{ka}\theta - \lambda\phi = 0, \\[4pt] \phi(\pm 1) = \phi'(\pm 1) = \theta(\pm 1) = 0, \end{array}\right\} \qquad (2.7)_k$$

for $k = 1, 2, \ldots$. We shall show that $(2.7)_k$ has a smallest, simple, eigenvalue. The proof of this fact requires some technicality and we will limit ourselves to giving the outline of the arguments, only. First one proves a lemma.

Lemma 2.1 *The operator M_{ka}^2 with the boundary conditions*

$$\phi(\pm 1) = \phi'(\pm 1) = 0$$

is invertible with compact positive inverse (that is maps positive functions into positive ones).

A proof of Lemma 2.1 can be found (in a more general form) in [**Ve**]. In addition, an elementary argument yields the following.

Lemma 2.2 *The operator M_{ka} with the boundary conditions*

$$\theta(\pm 1) = 0$$

is invertible with compact positive inverse.

Let $G = G_{ka}$ be the inverse of M_{ka} (with the boundary condition $\theta(\pm 1) = 0$) and let $H = H_{ka}$ be the inverse of M_{ka}^2 (with the boundary conditions $\phi(\pm 1) = \phi'(\pm 1) = 0$). Then $(2.7)_k$ is equivalent to

$$\theta = \lambda^2 GH(\theta).$$

Plainly, by regularity, we can work in $C([-1,1])$. Since G and H are compact and positive, an important result by Krasnoselski ([**Kr2**]), Chap.2,

Theorems 2.5 and 2.10) yields the existence of a unique eigenvalue with *positive* eigenfunction $\theta \geq 0$, and this eigenvalue is simple. The only assumption that remains to be verified is that GH is u_0-positive, that is, that there exist $u_0 \geq 0$, $u_0 \neq 0$, and a constant $\alpha > 0$, such that

$$GH u_0 \geq \alpha u_0.$$

For this, it suffices to set $u_0(y) = (y+1)^2(y-1)^2$ (or else any function positive in (0,1) with zero derivative at the endpoints); indeed, one shows readily that for all continuous nonnegative v, $v \neq 0$, the function Gv has positive (respectively, negative) first derivative at the point -1 (resp. 1).

In conclusion we can state the following result.

Lemma 2.3 *For each $k = 1, 2, \ldots$, problem $(2.7)_k$ has a smallest positive eigenvalue $\lambda(ka)$, which is simple.*

We remark that, if (2.7) has an eigenvalue λ with eigenfunction (ϕ, θ), then it has also an eigenvalue $-\lambda$, with eigenfunction $(\phi, -\theta)$.

In addition, we need the following estimate.

Lemma 2.4 *There exists a constant $C > 0$ such that $\lambda(ka) \geq Ck^2a^2$.*

Proof. Let us set

$$\mathcal{G}(\phi, \theta) = \int\limits_{-1}^{1} \phi(y)\theta(y)\mathrm{d}y,$$

and

$$\mathcal{F}(\phi, \theta) = \frac{1}{2} \int\limits_{-1}^{1} [\phi_{yy}^2 + 2k^2a^2\phi_y^2 + k^4a^4\phi^2 + \theta_y^2 + k^2a^2\theta^2]\mathrm{d}y.$$

System $(2.7)_k$ can be seen as the Euler–Lagrange equation

$$\operatorname{grad} \mathcal{F}(\phi, \theta) = \lambda \operatorname{grad} \mathcal{G}(\phi, \theta)$$

where λ plays the rôle of the Lagrange multiplier. As a consequence, one can readily see that the value $\mu(ka) = 1/\lambda(ka)$ has the variational characterization $\mu(ka) = \max\{\mathcal{G}(\phi, \theta) : \mathcal{F}(\phi, \theta) = 1\}$. Hence one has

$$\mu(ka) = \sup \frac{\mathcal{G}(\phi, \theta)}{\mathcal{F}(\phi, \theta)}.$$

Since, plainly,

$$\mathcal{G}(\phi, \theta) \leq \frac{1}{2} \int\limits_{-1}^{1} (\phi^2 + \theta^2)\mathrm{d}y$$

and

$$\mathcal{F}(\phi, \theta) \geq \frac{1}{2} k^2 a^2 \int\limits_{-1}^{1} (2\phi_y^2 + \theta^2) dy,$$

then one has

$$\frac{\mathcal{G}(\phi, \theta)}{\mathcal{F}(\phi, \theta)} \leq \frac{1}{k^2 a^2} \frac{\frac{1}{2}\int\limits_{-1}^{1}(\phi^2 + \theta^2)dy}{\int\limits_{-1}^{1}(2\phi_y^2 + \theta^2)dy},$$

From the Poincaré inequality we deduce that

$$\frac{\int\limits_{-1}^{1}(\phi^2 + \theta^2)dy}{\int\limits_{-1}^{1}(2\phi_y^2 + \theta^2)dy} \leq \text{ constant,}$$

and thus $\mu(ka) \leq C/k^2a^2$, proving the lemma.

We are now in position to go back to problem (2.5). Let R_a be such that

$$\sqrt{R_a} = \frac{\lambda(a)}{a}.$$

Let (ϕ_1^*, θ_1^*) denote an eigenfunction of $(2.7)_1$ corresponding to $\lambda(a)$. From Lemma 2.3 it follows that $\sqrt{R_a}$ is an eigenvalue of (2.5) with eigenfunction

$$\left.\begin{array}{l} \phi^*(x, y) = \phi_1^*(y) \cos ax, \\ \theta^*(x, y) = \theta_1^*(y) \sin ax. \end{array}\right\}$$

Suppose that

$$\frac{\lambda(a)}{a} < \frac{\lambda(ka)}{ka}, \text{ for all integers } k \geq 2. \qquad (2.8)$$

Since $\lambda(ka)/ka$ is the smallest positive eigenvalue of $(2.6)_k$, it is clear that $\lambda(a)/a$ is different from any eigenvalue of $(2.6)_k$ corresponding to an integer $k \geq 2$. If $(\phi, \theta) \in \text{Ker}(L_R)$ then the Fourier components ϕ_k, θ_k of ϕ, θ satisfy

$$\left.\begin{array}{l} M_{ka}^2 \phi_k - \lambda(a)k\theta_k = 0, \\ M_{ka}\theta_k - \lambda(a)k\phi_k = 0. \end{array}\right\}$$

But the preceding arguments show that $\lambda(a)k$ is not an eigenvalue of $(2.7)_k$ for $k \geq 2$. Hence $\phi_1 = \alpha\phi_1^*$, $\theta_1 = \theta_1^*$ and $\phi_k = \theta_k = 0$ for all $k \geq 2$. This proves that $\text{Ker}(L_R)$ is one-dimensional and spanned by (ϕ_1^*, θ_1^*). In this case Theorem 5.4.1 yields the existence of a branch of solutions of (2.2–3) bifurcating from $R = R_a$, $\psi = 0$, $\theta = 0$.

If (2.8) does not hold then from Lemma 2.4, there exists an integer k^* such that

$$\frac{\lambda(k^*a)}{k^*a} \le \frac{\lambda(ja)}{ja} \text{ for all } j > k^*. \tag{2.9}$$

Setting $a^* = k^*a$, from (2.9) we get that

$$\frac{\lambda(a^*)}{a^*} < \frac{\lambda(ka^*)}{ka^*} \text{ for all } k \ge 2,$$

which is the same as (2.8) with a replaced by a^*. It follows that there exists a bifurcation of solutions, with period $2\pi/a^*$, from R^*, $\sqrt{R^*} = \lambda(a^*)/a^*$. Since $2\pi/a^*$ is a sub-multiple of $2\pi/a$, these solutions are $(2\pi/a)$-periodic, too. In conclusion, we can state the following.

Theorem 2.5 *For any fixed $\tau > 0$, there exists $R_\tau > 0$ such that the system (2.2)–(2.3) possesses a family of solutions (ψ_R, θ_R), τ-periodic with respect to the x variable, depending continuously on the Rayleigh number R for R in a neighbourhood of R_τ. Moreover, as $R \to R_\tau$, $(\psi_R, \theta_R) \to (0,0)$ in $H_a^4 \times H_a^2$.*

3 Small oscillations for second-order dynamical systems

In some cases the problem to be studied inherits a specific symmetry and the eigenvalues of the linearized equation are not simple. However, it can still be possible to apply the bifurcation theorems of Section 5.4, working in suitable invariant subspaces. In this and in the next section we will discuss two such problems.

Here we will discuss a result due to Hopf [**Ho**] dealing with second-order *autonomous* systems of the type

$$\frac{d^2u}{dt^2} = f(u), \tag{S}$$

where $u \in \mathbb{R}^n$ and $f : \mathbb{R}^n \to \mathbb{R}^n$.

If $f(0) = 0$, (S) has the *trivial* solution $u \equiv 0$ and we look for periodic solutions of (S) "near" $u \equiv 0$.

To state in a more precise way the problem we want to address, it is convenient to put in evidence in (S) the dependence on the period. For this, let us perform a change of scale of time letting $s = \omega t$. Then (S) becomes

$$\omega^2 \frac{d^2u}{ds^2} = f(u). \tag{S_ω}$$

If, for some $\omega > 0$, $u = u(s)$ is a 2π-periodic solution of (S_ω), then

$$\tilde{u}(t) = u(\omega t)$$

is a T-periodic solution of (S), with $T = 2\pi/\omega$.

The problem of the existence of small oscillations near $u = 0$ can now be made more precise.

Let $\omega^* > 0$ be a value with the property that there exist

(a) a sequence $\omega_n \to \omega^*$,
(b) a sequence u_n of 2π-periodic solutions of (S_{ω_n}) such that $\|u_n\|_\infty \downarrow 0$.

Defining $\tilde{u}_n(t) = u_n(\omega_n t)$, we find that $\tilde{u}_n$ is a sequence of $(2\pi/\omega_n)$-periodic solutions of (S), with $\|\tilde{u}_n\|_\infty \downarrow 0$.

In conclusion, we can give the following definition.

Definition 3.1 If $\omega^* > 0$ is such that (a) and (b) hold true, then we will say that (S) possesses *small oscillations in correspondence with the frequency* ω^*.

Problem (S_ω) fits into the framework of bifurcation theory, the frequency ω playing the role of the bifurcation parameter. Roughly, we can explain this claim by noticing that the solutions (ω, u) of

$$F(\omega, u) := \omega^2 \frac{d^2 u}{ds^2} - f(u) = 0, \tag{3.1}$$

($\omega > 0$, u in a suitable space of smooth 2π-periodic solutions; see later on) correspond to solutions of (S_ω). There results $F(\omega, 0) \equiv 0$ and, moreover, if ω^* is a bifurcation point of $F = 0$, then, by definition, there exists a sequence $(\omega_n, u_n) \to (\omega^*, 0)$ such that $u_n \neq 0$ and $F(\omega_n, u_n) = 0$. This means that (ω_n, u_n) satisfy (a)–(b). Since the converse is also true, we can conclude that (S) possesses small oscillations in correspondence with the frequency ω^* if and only if ω^* is a bifurcation point of (3.1).

We will suppose

(f0) $f \in C^2(\mathbb{R}^n, \mathbb{R}^n)$, $f(0) = 0$,
(f1) $A := f'(0)$ is non-singular and has $r, r \geq 1$, negative real eigenvalues

$$-\omega_1^2, -\omega_2^2, \ldots, -\omega_r^2,$$

with, say, $0 < \omega_1 < \omega_2 < \ldots < \omega_r$.

Theorem 3.2 *Suppose f satisfies* (f0–1). *Let ω_j be such that*

(f2) $-\omega_j^2$ *is simple (i.e. has algebraic multiplicity = 1),*
(f3) ω_s/ω_j *is not an integer, for all $s \neq j$.*

Then (S) *possesses small oscillations in correspondence with the frequency ω_j.*

Proof. Let $C_{2\pi}^k$ denote the space of 2π-periodic functions of class $C^k(\mathbb{R}, \mathbb{R}^n)$, $C_{2\pi} = C_{2\pi}^0$, and let $F : \mathbb{R} \times C_{2\pi}^2 \to C_{2\pi}$ be defined by

$$F(\omega, u) = \omega^2 \frac{d^2 u}{ds^2} - f(u). \tag{3.2}$$

According to the preceding arguments, we want to show that ω_j is a bifurcation for F. This will be done using Theorem 5.4.1 with $\lambda^* = \omega_j$. For this purpose, we have to study the derivative $L = F_u(\omega_j, 0)$,

$$L : u \to \omega_j^2 \frac{d^2 u}{ds^2} - Au.$$

It is easy to see that L has a two-dimensional kernel. For example, if $n = 1$ and $A = -\omega_j^2$ then $\mathrm{Ker}(L)$ is spanned by $\cos t$ and $\sin t$. To overcome this difficulty, it is convenient to restrict F to the subspaces

$$X = \{x \in C_{2\pi}^2 : u(-t) = u(t)\},$$
$$Y = \{v \in C_{2\pi} : v(-t) = v(t)\}.$$

The restriction $F|_X$ will still be denoted by F. Let us note that $F(\omega, .)$ maps X into Y because (S) is an *autonomous system and does not contain u'.* Formally the Fourier expansion of any $u \in X, v \in Y$ is

$$u = u_0 + u_1 \cos t + u_2 \cos 2t + \ldots = \sum_{k \geq 0} u_k \cos kt,$$

$$v = v_0 + v_1 \cos t + v_2 \cos 2t + \ldots = \sum_{k \geq 0} v_k \cos kt,$$

with $u_k, v_k \in \mathbb{R}^n$.

For $k \in \mathbb{N}$ we set $\mathcal{M}_k = -k^2 \omega_j^2 I - A$ and $\mathcal{M} = \mathcal{M}_1 = -\omega_j^2 I - A$. By assumption $\mathrm{Ker}(\mathcal{M})$ is one-dimensional; we let $\xi \in \mathbb{R}^n$ denote a vector spanning $\mathrm{Ker}(\mathcal{M})$ and Π the corresponding spectral projector, namely the projector of $\mathbb{R}^n$,

$$x \to \xi \langle \eta, x \rangle = \Pi x,$$

where η is the functional that is zero on $R(\mathcal{M})$ and such that $\langle \eta, \xi \rangle = 1$.

We need the following

Lemma 3.3

(i) $\mathrm{Ker}(L) = \mathrm{span}\{\xi \cos t\}$,

(ii) $\mathrm{Range}(L) = \{v \in Y : \Pi v_1 = 0\}$ *where* $v_1 = \frac{1}{\pi} \int_0^{2\pi} v(t) \cos t \, dt$.

Proof. Consider the equation $Lu = v$, with $u \in X$ and $v \in Y$. If u_k, v_k denote the Fourier coefficients of u, v, than $Lu = v$ becomes

$$-k^2 \omega_j^2 u_k - A u_k \equiv \mathcal{M}_k(u_k) = v_k, \quad (k \in \mathbb{N}). \tag{3.3$_k$}$$

If $Lu = v$ then u_k, v_k satisfy $(3.3)_k$; conversely, if $u \in X$ and $v \in Y$ and u_k, v_k satisfy $(3.3)_k$ then $Lu = v$.

We claim that (3.3) has a unique solution for all $k \neq 1$. In fact, if $k = 0$ then $\mathcal{M}_0$ is nothing but A, which is invertible; if $k \neq 1$ and $k \neq 0$, then from (f2) it follows that $-k^2\omega_j^2$ is not an eigenvalue of A and hence $\mathcal{M}_k(k \neq 0, 1)$ is still invertible. In conclusion $(3.3)_{k \neq 1}$ has a unique solution given by

$$u_k = \mathcal{M}_k^{-1}(v_k) \quad (k \neq 1).$$

For $k = 1$ (3.3) becomes

$$\omega_j^2 u_1 - Au_1 \equiv \mathcal{M}u_1 = v_1,$$

which is solvable whenever $\Pi v_1 = 0$.

After these preliminares we can now prove (i) and (ii).

First, if $v = 0$ then the solutions of (3.3) are given by

$$\left. \begin{array}{l} u_k = 0 \text{ if } k \neq 1, \\ u_1 = a\xi \; (a \in \mathbb{R}). \end{array} \right\}$$

Hence one has $\mathrm{Ker}(L) = \mathrm{span}\{\xi \, \cos t\}$. This proves (i).

To prove (ii), we first let $v = Lu$, for some $u \in X$. The preceding arguments imply that $\Pi v_1 = 0$ and hence $\{v \in Y : \Pi v_1 = 0\} \supset R = \mathrm{Range}(L)$. Conversely, let $v \in Y$ be such that $\Pi v_1 = 0$. Then, formally, $Lu = v$ has a solution of the form

$$u(t) = A^{-1}v_0 + u_1 \cos t + \tilde{v}(t) \tag{3.4}$$

where $\mathcal{M}(u_1) = v_1$ and

$$\tilde{v}(t) = \sum_{k \geq 2} \mathcal{M}_k^{-1}(v_k) \cos kt.$$

We claim that $\tilde{v}$ is of class C^2. To see this, we first notice that

$$\mathcal{M}_k^{-1} = (-k^2\omega_j^2 I - A)^{-1} = \frac{-1}{k^2\omega_j^2}(I + \frac{1}{k^2\omega_j^2}A)^{-1}$$

$$= \frac{-1}{k^2\omega_j^2}I + \mathcal{N}_k,$$

where

$$\mathcal{N}_k(x) = O(\frac{1}{k^4}) \text{ for all } x \in \mathbb{R}^n \tag{3.5}$$

Then it follows that

$$\tilde{v}(t) = w(t) + z(t)$$

with

$$w(t) = \sum_{k \geq 2} -\frac{1}{k^2\omega_j^2}v_k \cos kt$$

and

$$z(t) = \sum_{k \geq 2} \mathcal{N}_k(v_k) \cos kt.$$

Since $w(t)$ is nothing but the iterated integral of the continuous function

$$\frac{1}{\omega_j^2} \sum_{k \geq 2} v_k \cos kt,$$

then $w \in C^2$.

As for $z(t)$, we remark that, v being continuous, (3.5) yields

$$\mathcal{N}_k(v_k) = o(\frac{1}{k^4}).$$

Hence z is also of class C^2. In conclusion, u given by (3.4) is C^2 and is a solution of $Lu = v$. This completes the proof of (ii).

Proof of Theorem 3.2 completed.   From Lemma 3.3 it follows that assumption (I) of Theorem 5.4.1. holds true. As for hypothesis (4.3), it suffices to notice that here

$$F_{u,\lambda}(\omega_j, 0) : u \to 2\omega_j \frac{\mathrm{d}^2 u}{\mathrm{d}t^2}$$

and hence (1.2) is satisfied too. Then Theorem 5.4.1 applies and the result follows.

4 Water waves

In this section we will discuss a bifurcation result dealing with the classical problem of the existence of periodic waves of a heavy, inviscid fluid with infinite depth. One of the first results on this matter is due to Levi–Cività [**LC**], and Krasovski (see [**Be**] for an outline of the Krasovski result. Periodic water waves with infinite depth have been studied by Nekrasov [**Ne**]). For other problems concerning waves in fluid dynamics, see, for example [**T**].

Let us consider a two-dimensional fluid with uniform density $\rho = 1$, say and infinite depth, having a free bounded surface C. We will take the x-axis in the horizontal direction and the y-axis upward directed. We will look for waves moving with constant velocity $c > 0$. It is convenient to take a reference frame which is in translational motion with velocity c. Then the profile of the wave becomes independent of time and the velocity field $\mathbf{q} = (u, v)$ will be a function of x and y only. Moreover, since the fluid is assumed to be at rest at great depth, it will move with velocity $-c$ with respect to the new reference frame, and this leads to

$$\mathbf{q} \to (-c, 0) \text{ as } y \to -\infty, \text{ uniformly with respect to } x. \qquad (4.1)$$

Let us also assume that C has equation $y = \beta(x)$.

The fluid is assumed to be irrotational and incompressible so that there exist two functions $\varphi(x, y)$ (velocity potential) and $\psi(x, y)$ (stream function) such that

$$\nabla\varphi = (u, v), \ \nabla\psi = (-v, u).$$

Such φ and ψ are conjugate harmonic functions.

Let us introduce the following notation:

$$z = x + iy,$$

$$w = u - iv,$$

$$f = \varphi + i\psi,$$

where i is the imaginary unit ($i^2 = -1$). There result

$$w^* = u + iv \ \text{ and } \ |w^2| = u^2 + v^2 = |\mathbf{q}|^2.$$

Moreover, the profile C of the wave is a streamline $\psi =$const. and we will assume that

$$\psi(x, y) = 0 \text{ for all } (x, y) \in C.$$

The Bernoulli law yields (recall that $\rho = 1$)

$$\frac{1}{2}|w|^2 + gy + p = \text{const.}, \tag{4.2}$$

where g is the gravitational constant and p is the pressure. Since p is also constant on the upper surface C (actually p is the atmospheric pressure), (4.2) implies

$$\frac{1}{2}|w|^2 + gy = \text{const. on } C. \tag{4.3}$$

We will look for periodic waves, that is for velocity fields that are periodic with respect to x with period $h \in \mathbb{R}$; in terms of w this means we are looking for w satisfying

$$w(z + h) = w(z) \text{ for all } z \in \mathbb{C} \tag{4.4}$$

for some (unknown) real h.

As a consequence the holomorphic function f has derivative $df/dz = w$ which is h-periodic. Hence from (4.4) it follows that

$$f(z + h) - f(z) = \text{const.} \tag{4.5}$$

To find the value of this constant, it suffices to put $z = iy$ and let $y \to -\infty$ in (4.5), using the integral representation formula of $\varphi + i\psi$. So there results

$$f(z + h) - f(z) = ch. \tag{4.6}$$

The above problem is a free boundary one, in the sense that the region filled by the fluid $A = \{y \le \beta(x)\}$ is unknown. To overcome this

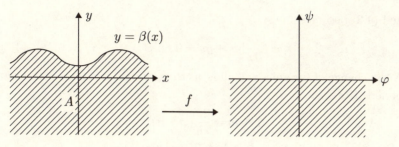

Figure 6.1

difficulty one looks for a (holomorphic) transformation of coordinates so that such a region becomes a fixed domain. To this purpose, it is convenient to use the same $f(z)$. This choice is suitable for a change of coordinates provided $|f'(z)| = |w| \geq$ const. > 0 for all $z \in A$. If this is the case, the map $z \to f$ is globally invertible with inverse denoted by Φ. The region A is transformed into the region $\{f = \varphi + i\psi : \psi \leq 0\}$. (See Figure 6.1.)

The function w is holomorphic with respect to z and therefore with respect to f. As a function of f, w is periodic with period ch. To see this fact, we start from (4.6), that is $f(z + h) = f(z) + ch$, and apply Φ to find

$$\Phi(f + ch) = z + h = \Phi(f) + h.$$

This implies $\Phi'(f + ch) = \Phi'(f)$. Since $\Phi' = dz/df = 1/w$, we conclude that

$$w(f + ch) = w(f)$$

as claimed.

It remains to translate the equation (4.3). To do that, it suffices to note that one has

$$\frac{1}{2}|w|^2 + gy = \text{const. on } \psi = 0.$$

Differentiating with respect to φ we find

$$\frac{1}{2}\frac{\partial}{\partial \varphi}|w|^2 + g\frac{\partial y}{\partial \varphi} = 0 \text{ on } \psi = 0.$$

Note that $\partial y/\partial \varphi$ is the imaginary part of $\partial z/\partial \varphi$. Since there results

$$\frac{\partial z}{\partial \varphi} = \frac{dz}{df} = \frac{1}{df/dz} = \frac{1}{u - iv} = \frac{u + iv}{u^2 + v^2},$$

it follows that

$$\frac{\partial y}{\partial \varphi} = \frac{v}{|w|^2},$$

and (4.3) becomes

$$|w|^3 \frac{\partial}{\partial \varphi} |w| + gv = 0 \text{ on } \psi = 0. \tag{4.7}$$

A last change of variables will allow us to work with functions with fixed period. If we set

$$\xi + i\eta = \frac{2\pi}{ch} (\varphi + i\psi),$$

w becomes 2π-periodic in ξ and (4.7) transforms into

$$|w|^3 \frac{\partial}{\partial \xi} |w| + \frac{gch}{2\pi} v = 0 \text{ on } \eta = 0. \tag{4.8}$$

In conclusion we can formulate our problem in the following way:

Find a function $w = u - iv$ of the variable $\xi + i\eta$, such that

(i)   *w is holomorphic in the half-plane $\Omega = \{\xi + i\eta : \eta \leq 0\}$,*
(ii)   *w is 2π-periodic with respect to ξ,*
(iii) *w satisfies (4.8) on $\eta = 0$, and*
(iv)   *$w \to c$ as $\eta \to -\infty$, uniformly in ξ.*

In other words we will look for functions u, v defined in Ω, such that

(i′)   *u and v are harmonic conjugate functions,*
(ii′) *u and v are 2π-periodic with respect to ξ,*
(iii′) *u and v satisfy (4.8) on $\eta = 0$,*
(iv′) *as $\eta \to -\infty$, $u \to c$ and $v \to 0$.*

To translate the preceding conditions into a functional equation, some preliminaries are in order.

Let E denote the class of real-valued functions u defined on the half-plane Ω such that (i) u is harmonic in Ω, (ii) u is 2π-periodic with respect to ξ, (iii) the function $\xi \to u(\xi, \eta)$ is bounded in $L^2(0, 2\pi)$ for $\eta \in (-\infty, 0)$.

Lemma 4.1   *Any $u \in E$ has the form*

$$u(\xi, \eta) = \gamma + \sum_{k \geq 1} (a_k \cos k\xi + b_k \sin k\xi) e^{k\eta},$$

where $\gamma \in \mathbb{R}$ and $\sum (a_k^2 + b_k^2) < \infty$.

Proof. For fixed $\eta < 0$, the function $u(., \eta)$ has Fourier expansion

$$u(\xi, \eta) = \ell(\eta) + \sum_k (\alpha_k(\eta) \cos k\xi + \beta_k(\eta) \sin k\xi).$$

Since u is harmonic,

$$\left.\begin{aligned}
\ell''(\eta) &= 0, \\
\alpha_k''(\eta) &= k^2 \alpha_k(\eta), \\
\beta_k''(\eta) &= k^2 \beta_k(\eta),
\end{aligned}\right\}$$

whence

$$\left.\begin{aligned}
\ell(\eta) &= m\eta + \ell_0, \\
\alpha_k(\eta) &= a_k e^{k\eta} + \hat{a}_k e^{-k\eta}, \\
\beta_k(\eta) &= b_k e^{k\eta} + \hat{b}_k e^{-k\eta}.
\end{aligned}\right\}$$

Since $u(.,\eta)$ is bounded in $L^2(0, 2\pi)$ as $\eta \to -\infty$, one has

$$m = 0, \quad \hat{a}_k = 0, \quad \hat{b}_k = 0,$$

as well as $\ell_0 = \gamma$ and

$$\Sigma(a_k^2 + b_k^2)e^{k\eta} \le L.$$

From this last inequality it follows readily that $\Sigma(a_k^2 + b_k^2) \le L$, as claimed.

Remark 4.2   From Lemma 4.1 we infer that any $u \in E$ has, in addition to (i)–(ii), (iii), the following properties:

(iv)   *as $\eta \to -\infty$, $u(\xi, \eta) \to \gamma$, uniformly in ξ,*
(v)   *u has a trace on $\eta = 0$, denoted by $u_\# = u_\#(\xi)$, which is the L^2-limit of $u(.,\eta)$ as $\eta \to 0$.*

It is worth noticing that γ is nothing but the mean value of $u_\#$. In the sequel the mean value of a function $z \in L^2(0, 2\pi)$ will be denoted by $[z]$.

Let us introduce the following function space:

$$H = \{u_0 \in L^2(\mathbb{R}) : u_0 \text{ is } 2\pi\text{–periodic and } [u_0] = 0\}.$$

With this notation, for any $u \in E$ one has that $u_\# = \gamma + u_0$, with $u_0 \in H$.

The solution u we are seeking is therefore a function in E of the form

$$u(\xi, \eta) = c + \sum_{k \ge 1}(a_k \cos k\xi + b_k \sin k\xi)e^{k\eta} \tag{4.9}$$

and its trace on $\eta = 0$ is given by $u_\# = c + u_0$, with $u_0 \in H$.

The other unknown v is such that $-v$ is the harmonic conjugate of u, and $v \to 0$ as $\eta \to -\infty$, uniformly on ξ. Hence, if u is given by (4.9) then v will have the form

$$v(\xi, \eta) = \sum_{k \ge 1}(-b_k \cos k\xi + a_k \sin k\xi)e^{k\eta}.$$

The trace of such a v has mean value equal to zero and therefore belongs to H. It is convenient to introduce a map $K : H \to H$ in the following way: if $z \in H$ is of the form $\sum_{k \geq 1} (p_k \cos k\xi + q_k \sin k\xi)$ then Kz is given by

$$Kz = \sum_{k \geq 1} (-q_k \cos k\xi + p_k \sin k\xi). \qquad (4.10)$$

Plainly, K is a linear continuous map from H into itself (indeed, an isometry).

According to the preceding remarks, the solutions of our problem will be u and v given by

$$u(\xi, \eta) = c + \sum_{k \geq 1} (a_k \cos k\xi + b_k \sin k\xi) e^{k\eta},$$

$$v(\xi, \eta) = \sum_{k \geq 1} (-b_k \cos k\xi + a_k \sin k\xi) e^{k\eta}.$$

The trace of u on $\eta = 0$ will be of the form $c + u_0$, with $u_0 \in H$, while that of v will be $v_0 = K(u_0)$. With this notation there results

$$|w|^2 = (c + u_0)^2 + v_0^2 = (c + u_0)^2 + (Ku_0)^2 \text{ on } \eta = 0. \qquad (4.11)$$

To write equation (4.8) in terms of u_0 and v_0 we shall introduce some more notation: for $z \in L^2$, we indicate with Πz the primitive of z such that $[\Pi z] = 0$.

In other words, if

$$z(\xi) = \sum_{k \geq 1} (p_k \cos k\xi + q_k \sin k\xi)$$

one has

$$\Pi z(\xi) = \sum_{k \geq 1} \left(-\frac{q_k}{k} \cos k\xi + \frac{p_k}{k} \sin k\xi \right),$$

and, by (4.10),

$$\Pi K z(\xi) = \sum_{k \geq 1} \left(-\frac{p_k}{k} \cos k\xi - \frac{q_k}{k} \sin k\xi \right). \qquad (4.12)$$

After these preliminaries, let us integrate (4.8) with respect to ξ; one finds

$$\frac{1}{4} |w|^4 + \frac{gch}{2\pi} \Pi v = \text{const. on } \eta = 0. \qquad (4.13)$$

Taking into account (4.11), and dividing by c^3, we find that (4.13) becomes

$$\frac{1}{4c^3} \left((c + u_0)^2 + (Ku_0)^2 \right)^2 + \lambda \Pi K u_0 = \text{const.}, \lambda = \frac{gh}{2\pi c^2}.$$

Taking the mean value of both sides we find that the constant in the right-hand side is nothing but $(1/4c^3)[((c + u_0)^2 + (Ku_0)^2)^2]$. In conclusion, letting $F : H \to H$,

$$F(\lambda, u_0) = \frac{1}{4c^3}((c+u_0)^2 + (Ku_0)^2)^2 - \frac{1}{4c^3}\left[((c+u_0)^2 + (Ku_0)^2)^2\right] + \lambda \Pi K u_0,$$

we shall solve $F(\lambda, u_0) = 0$.

Plainly, F is smooth (C^∞) and $F(\lambda, 0) \equiv 0$. The derivative $F_{u_0}(\lambda, 0)$ is the map

$$z \to z + \lambda \Pi K z.$$

Taking into account (4.12) we get that

$$z = \Sigma_{k \geq 1} \, (p_k \cos k\xi + q_k \sin k\xi) \in \text{Ker}(F_{u_0}(\lambda, 0))$$

whenever

$$\left. \begin{array}{l} p_k - \lambda \dfrac{p_k}{k} = 0, \\[2mm] q_k - \lambda \dfrac{q_k}{k} = 0 \end{array} \right\} \quad (k = 1, 2, \ldots).$$

As a consequence, $\text{Ker}(F_{u_0}(\lambda, 0)) \neq \{0\}$ if and only if $\lambda = 1, 2, \ldots$. Moreover, for $\lambda = \kappa \in \mathbf{N}$, such a kernel is two-dimensional and spanned by $(\cos \kappa\xi, \sin \kappa\xi)$. In view of this fact, we can proceed as in §3 and consider the subspace X of H consisting of those z that are even. The restriction $F^* = F|_X$ maps X into itself, because $\Pi K z$ is an even function provided z is. Let $L_\kappa = F^*_{u_0}(\kappa, 0)$. According to the preceding discussion it is easy to check that, for each $\kappa = 1, 2, \ldots$, $\text{Ker}(L_\kappa)$ is one-dimensional and $\text{R}(L_\kappa)$ has codimension 1. Moreover the second mixed partial derivative $F^*_{u_0, \lambda}(\kappa, 0)$ is the map $z \to \Pi K z$ and hence assumption (4.3) of Theorem 5.4.1 holds true as well.

In conclusion, we can state the following.

Theorem 4.3 *For all $\kappa = 1, 2, \ldots$ there is a branch of nontrivial solutions of $F(\lambda, u_0) = 0$ bifurcating from $(\kappa, 0)$. Moreover, each nontrivial solution (λ, u_0) gives rise to a solution $w = u - iv$ of the problem (4.8').*

Theorem 4.3 can be completed by some remarks. We shall be sketchy here, leaving the details to the reader.

According to Remark 5.4.3(iv) one can verify that the branch bifurcating from $(\kappa, 0)$, parametrized with respect to μ, has equation

$$u_{0, \kappa}(\mu, \xi) = \mu \cos \kappa\xi + \omega_\kappa(\mu, \xi),$$

$$\lambda_\kappa(\mu) = \kappa + \gamma_\kappa(\mu),$$

where $\omega_\kappa(\mu, .)$ is even, $\omega_\kappa = O(\mu^2)$ as $\mu \to 0$, uniformly with respect to ξ, and γ_κ is even. Of course, the corresponding $v_{0, \kappa}$ has the form

$$v_{0, \kappa}(\mu, \xi) = \mu \sin \kappa\xi + \chi_\kappa(\mu, \xi),$$

where $\chi_\kappa(\mu, .)$ is odd and $\chi_\kappa = O(\mu^2)$ as $\mu \to 0$.

It can be also checked, that for $\kappa = 1$, there results

$$\gamma_1(\mu) = -\frac{\mu^2}{c^2} + O(\mu^4).$$

Moreover, let us point out that the solutions on the branch Γ_κ emanating from $(\kappa, 0)$ correspond to the same wave profile as that of the solutions on the first branch Γ_1. To make this claim more precise, consider the solutions bifurcating from $(1, 0)$,

$$u_{0,1}(\mu, \xi) = \mu \cos \xi + \omega_1(\mu, \xi),$$

$$\lambda_1(\mu) = 1 + \gamma_1(\mu),$$

that satisfy

$$F(\lambda_1(\mu), u_{0,1}) = 0.$$

Letting

$$\hat{u}_{0,\kappa}(\mu, \xi) = u_{0,1}(\mu, k\xi),$$

$$\hat{\lambda}_\kappa(\mu) = \kappa + \kappa\gamma_1(\mu),$$

one can immediately verify that there results

$$F(\hat{\lambda}_\kappa(\mu), \hat{u}_{0,\kappa}) = 0.$$

Since Γ_κ is the unique branch of nontrivial solutions bifurcating from $(\kappa, 0)$, it follows that

$$u_{0,\kappa} = \hat{u}_{0,\kappa} = u_{0,1}(\mu, \kappa\xi),$$

$$\lambda_\kappa(\mu) = \hat{\lambda}_\kappa(\mu) = \kappa + \kappa\gamma_1(\mu).$$

Hence the solutions on Γ_κ are given by

$$u_{0,\kappa}(\mu, \xi) = \mu \cos \kappa\xi + \omega_1(\mu, \kappa\xi),$$

$$\lambda_\kappa(\mu) = \kappa + \kappa\gamma_1(\mu),$$

and therefore they give rise to the same wave profile as that corresponding to the solutions on Γ_1.

5 Periodic solutions of a semilinear hyperbolic equation

This section deals with the existence of periodic solutions of a class of semilinear wave equations. We anticipate that the problem we are going to discuss cannot be handled directly by the abstract results of Section 5.4; however, it fits into that general framework and will be solved by means of a suitable Lyapunov–Schmidt reduction.

Let us consider the strip $S = \{(x,t) \in \mathbb{R}^2 : x \in [0, \pi], -\infty < t < +\infty\}$ and a function $f : S \times \mathbb{R} \to \mathbb{R}$, such that

$$f(x, t + 2\pi, u) = f(x, t, u) \text{ for all } (x, t, u) \in S \times \mathbb{R}. \qquad (5.1)$$

We will look for solutions $u = u(x,t)$, $(x,t) \in S$, of the following semilinear hyperbolic problem

$$\left.\begin{array}{l} \mathcal{A}u + \varepsilon f(x,t,u) = 0 \text{ for all } (x,t) \in S, \\ u(0,t) = u(\pi,t) = 0 \text{ for all } t \in \mathbb{R} \\ u(x,t+2\pi) = u(x,t) \text{ for all } (x,t) \in S, \end{array}\right\} \qquad (5.2)$$

where ε is a real parameter and $\mathcal{A}$ denotes the wave operator:

$$\mathcal{A} = \frac{\partial^2}{\partial t^2} - \frac{\partial^2}{\partial x^2}.$$

For $\varepsilon = 0$ (5.2) has the trivial solution $u \equiv 0$ and our goal will be to show that, under suitable assumptions on the nonlinear term f, (5.2) has a unique nontrivial solution for all $\varepsilon \neq 0$ sufficiently small.

A result of this sort was first proved by P. H. Rabinowitz [**R1**]. Subsequently, L. De Simon & G. Torelli [**DST**] gave a new proof of this existence result. The arguments of [**DST**] not only are very simple, but also point out clearly the bifurcation framework that is beyond the problem. We shall expound only some of the material contained in [**DST**], referring to that paper for more details as well as for additional results.

Let H denote the space of real L^2 functions $u : S \to \mathbb{R}, 2\pi$-periodic in time. With respect to the usual scalar product

$$(u|v) = \int_0^\pi \mathrm{d}x \int_0^{2\pi} u(x,t)v(x,t)\mathrm{d}t$$

H is a Hilbert space. We will set $\|u\|^2 = (u|u)$.

We also indicate by $C_{2\pi}^k$ the space of functions $u \in C^k(S)$ that are 2π-periodic with respect to t; we will set $C_{2\pi}$ for $C_{2\pi}^0$.

Definition 5.1 Let $g \in H$. By a *generalized solution* of

$$\mathcal{A}u = g, \ u(0,t) = u(\pi,t) = 0 \text{ for all } t \in \mathbb{R} \qquad (5.3)$$

we mean a $u \in H$ satisfying

$$(u|\mathcal{A}\phi) = (g|\phi) \text{ for all } \phi \in C_{2\pi}^2, \ \phi = 0 \text{ on } \partial S. \qquad (5.4)$$

Let us notice that, since the x-partial derivatives of the test function ϕ take arbitrary values on ∂S, any $u \in C_{2\pi}^2$ satisfying (5.4) is zero on ∂S. In other words, the integral relationship (5.4) translates in the generalized sense also the boundary condition $u(0,t) = u(\pi,t) = 0$.

Here too, we will use complex notation for the sake of simplicity in computation. For $v \in H$ we will let $v_{m,n}$ denote the Fourier coefficients of v with respect to the orthonormal basis $\{\varphi_{m,n} = \sin mx \ \mathrm{e}^{int}\}, m = 1,2,\ldots,n = 0,\pm 1 \pm 2,\ldots$:

$$v = \sum_{m,n} v_{m,n} \sin mx \ \mathrm{e}^{int}.$$

The following lemma shows the relationship between a generalized solution of (5.3) and its Fourier coefficients.

Lemma 5.2   *A function $u \in H$ is a generalized solution of* (5.3) *if and only if there results*

$$(m^2 - n^2)u_{m,n} = g_{m,n} \text{ for all integers } m, n, \text{ with } m \geq 1 \qquad (5.5)$$

Proof. Let u be a generalized solution of (5.3). Taking as test functions $\phi = \varphi_{m,n}$, we find readily the relationships (5.5); conversely, if (5.5) holds, then by easy calculations one checks that $u = \Sigma u_{m,n}\varphi_{m,n}$ satisfies (5.4).

As an immediate consequence of Lemma 5.2 we find the following.

(a) If we denote by V the set of generalized solutions of $Lu = 0$, then V is a closed subspace of H and there results

$$V = \{v \in H : v_{m,n} = 0 \text{ for all } m^2 \neq n^2\}.$$

The projection onto V will be denoted by P.

(b) If W denotes the orthogonal complement of V in H, then there results

$$W = \{w \in H : w_{m,n} = 0 \text{ for all } m^2 = n^2\}.$$

The projection onto W will be denoted by Q. Any $u \in H$ can be written in the form

$$u = v + w, \text{ with } v = Pu \text{ and } w = Qu.$$

From (5.5) it follows immediately that (5.3) has a solution if and only if $g \in W$. Moreover, if $g \in W$ then there is a *unique* $w \in W$ such that $\mathcal{A}w = g$. In other words, $\mathcal{A}$ is invertible on W with linear continuous inverse $G : W \rightarrow W$ satisfying

$$G(g) = w \Leftrightarrow \mathcal{A}w = g \ (g, w \in W).$$

Finally, if f satisfies the condition

(f4) $f(x, t, s)$ *is square-integrable in S for all $s \in \mathbb{R}$, and 2π-periodic in t*

then f induces a continuous Nemitski operator $f : H \rightarrow H$ in the usual way,

$$f(u)(x, t) = f(x, t, u(x, t)).$$

With this notation the generalized formulation of problem (5.1) becomes

$$\mathcal{A}u + \varepsilon f(u) = 0, \quad u \in H. \qquad (5.6)$$

Let us substitute $u = v + w$ into (5.6) and consider the equation

$$F(\varepsilon, v, w) \equiv \mathcal{A}w + \varepsilon f(v + w) = 0.$$

To frame this problem as one of bifurcation, it is convenient to take $v \in V$ as the "bifurcation parameter" and $(\varepsilon, w) \in \mathbb{R} \times W$ as the "unknowns". Indeed, one has $F(0, v, 0) = 0$, for all $v \in V$ and hence the trivial solutions are now given by $(\varepsilon, w) = (0, 0)$. So the problem turns out to be that of seeking the possible $v_0 \in V$ such that $F = 0$ has solutions (ε, v, w) near to $(0, v_0, 0)$ with $(\varepsilon, w) \neq (0, 0)$. Any such solution corresponds to a "non- trivial" solution of (5.6).

The main result of this section is contained in the following theorem.

Theorem 5.3   *In addition to* (f4) *we suppose*

(f5)   *there exist constants $k > h > 0$ such that $h(s_2 - s_1) \leq f(x, t, s_2) - f(x, t, s_1) \leq k(s_2 - s_1)$, for all $(x, t) \in S$, for all $s_1 < s_2 \in \mathbb{R}$. Then for all $\varepsilon \neq 0$ small enough* (5.1) *has a unique generalized solution.*

Proof. If we let $u = v + w (= Pu + Qu)$ (5.6) becomes

$$\mathcal{A}w + \varepsilon f(v + w) = 0.$$

Applying the projections P and Q we find the equivalent system

$$\left. \begin{array}{l} \varepsilon P[f(v + w)] = 0, \\ \mathcal{A}w + \varepsilon Q[f(v + w)] = 0. \end{array} \right\}$$

Hence, for $\varepsilon \neq 0$ one is led to

$$\left. \begin{array}{l} P[f(v + w)] = 0, \\ w + \varepsilon G Q[f(v + w)] = 0 \end{array} \right\} \tag{5.7}$$

We claim as follows.

Lemma 5.4   *For all $w \in W$ the equation $P[f(v+w)] = 0$ has a unique solution $v = K(w)$. Moreover there results*

$$\|K(w_2) - K(w_1)\| \leq \frac{k}{h} \|w_2 - w_1\|. \tag{5.8}$$

Proof of the lemma. Fixing $w \in W$ let us consider the map $M_w : V \to V$

defined by $M_w(v) = P[f(v + w)]$. For any $v', v'' \in V$ we have
$$(M_w(v'') - M_w(v')|v'' - v')$$
$$= (P[f(v'' + w)] - P[f(v' + w)]|v'' - v')$$
$$= (f(v'' + w) - f(v' + w)|v'' - v')$$
$$= \int_0^\pi dx \int_0^{2\pi} [f(x, t, v''(x, t) + w(x, t))$$
$$- f(x, t, v'(x, t) + w(x, t))][v''(x, t) - v'(x, t)]dt.$$

Using (f5) we readily obtain
$$(M_w(v'') - M_w(v')|v'' - v') \geq h\|v'' - v'\|^2$$

From a well-known lemma by G. Minty [**Mi**] on monotone operators it follows that $M_w(v) = 0$ has a unique solution $v = K(w)$. To prove that K is Lipschitz-continuous with constant $= k/h$, we let $v_i = K(w_i)$, $i = 1, 2$. From the definition of K one has $P[f(v_i + w_i)] = 0$ $(i = 1, 2)$, and hence
$$0 = \left|(P[f(v_2 + w_2)] - P[f(v_1 + w_1)] \mid v_2 - v_1)\right|$$
$$\geq \left|(f(v_2 + w_2) - f(v_1 + w_2)|v_2 - v_1)\right| \tag{5.9}$$
$$- \left|(f(v_1 + w_2) - f(v_1 + w_1)|v_2 - v_1)\right|.$$

Now, as in the preceding step, $(f5)$ implies
$$\left|(f(v_2 + w_2) - f(v_1 + w_2)|v_2 - v_1)\right| \geq h\|v_2 - v_1\|^2 \tag{5.10}$$
as well as
$$\left|(f(v_1 + w_2) - f(v_1 + w_1)|v_2 - v_1)\right| \leq k\|w_2 - w_1\| \|v_2 - v_1\|. \tag{5.11}$$
From (5.9)–(5.11) we deduce
$$h\|v_2 - v_1\|^2 \leq k\|w_2 - w_1\| \|v_2 - v_1\|,$$
whence (5.8).

Proof of Theorem 5.3 completed. System (5.7) is equivalent to
$$\left.\begin{array}{l} v = K(w), \\ w + \varepsilon GQ[f(K(w) + w)] = 0. \end{array}\right\}$$

The operator $w \to GQ[f(K(w) + w)]$ is obtained by composition of Lipschitz maps and is therefore Lipschitzian itself. Then for $|\varepsilon| \, (> 0)$ small enough such an operator is a contraction, and the equation $w + \varepsilon GQ[f(Kw + w)] = 0$ has a unique solution. This proves the theorem.

Remark 5.5 The proof of Theorem 5.3 shows that the solution of (5.1) depends continuously on $\varepsilon(\neq 0)$. As $\varepsilon \to 0$ one has that $w \to 0$ while v tends to the unique solution of $P(f(v)) = 0$, namely $v_0 = K(0)$.

According to the discussion preceding the statement of Theorem 5.3, this means that the "bifurcation set" for (5.6) branches off from $v_0 = K(0)$.

Remark 5.6 We have used a generalized formulation also for the boundary conditions on ∂S. It is easy to see that, if u is a generalized solution of (5.1), in the sense of Definition 5.1, then the function $u(\delta, .)$ tends to zero in $L^2(0, 2\pi)$ as $\delta \to 0$ or π. For details, see [**DST**].

7

Bifurcation of periodic solutions

This chapter is devoted to studying the bifurcation of periodic solutions of first-order autonomous systems. We shall see that the natural setting for this kind of problems is the so-called "Hopf bifurcation" which is concerned with mappings depending on a two-dimensional parameter. We will show that a suitable use of the Lyapunov–Schmidt reduction allows us to obtain the Hopf bifurcation in a rather straight way. Applications to the classical Lyapunov Centre Theorem and to the restricted three-body problem will also be given.

1 The Hopf bifurcation

Motivation

In this section we will investigate the existence of bifurcations for mappings depending on a two-dimensional parameter. One of the motivations of this study is the search for periodic solutions of autonomous systems such as

$$\frac{\mathrm{d}u}{\mathrm{d}t} = f(\mu, u) \qquad (\mathrm{S}_\mu)$$

depending on a real parameter μ. Since (S_μ) is autonomous, the period of the solutions we are looking for is *a priori* unknown, while in order to apply the methods of nonlinear functional analysis it is convenient to work in a space of functions with a fixed period. For this one makes the time rescaling $t \to t/\omega$ ($\omega > 0$), looking for 2π-periodic solutions of

$$\omega \frac{du}{dt} = f(\mu, u). \qquad (S_{\mu,\omega})$$

If u is a 2π-periodic solution of $(S_{\mu,\omega})$ then $\tilde{u}(t) = u(\omega t)$ is a $(2\pi/\omega$-periodic solution of (S_μ). Solutions of $(S_{\mu,\omega})$ correspond to solutions of a functional equation $F = 0$, depending upon the parameter pair $(\omega, \mu) \in \mathbb{R}^2$.

Studying the *bifurcation of periodic solutions* for (S_μ) we will deal with the case when $f(\mu, 0) \equiv 0$; then $(S_{\mu,\omega})$ has the *trivial* solution $u \equiv 0$ and we are interested in finding the possible *bifurcation* values (ω_0, μ_0) where *nontrivial* periodic solutions originate.

The Abstract Bifurcation Theorem

Motivated by the preceding considerations, which will be discussed in more detail in the following sections, we consider here from the abstract point of view a Banach space X and a map $F : \mathbb{R}^2 \times X \to Y$ such that

$$F \in C^2(\mathbb{R}^2 \times X, Y) \text{ and } F(\omega, \mu, 0) = 0 \text{ for all } (\omega, \mu) \in \mathbb{R}^2, \qquad (1.1)$$

in such a way that the equation $F(\omega, \mu, u) = 0$ has the trivial solution $u = 0$ for all $(\omega, \mu) \in \mathbb{R}^2$. Extending in a natural way the definition given in Section 5.1, we will say that

(ω_0, μ_0) is a bifurcation point for $F = 0$ if there exists a sequence

$$(\omega_n, \mu_n) \to (\omega_0, \mu_0)$$

and a sequence $u_n \neq 0$ such that $F(\omega_n, \mu_n, u_n) = 0$.

It is rather natural to expect that the situation corresponding to the "bifurcation from the simple eigenvalue" will now be concerned with the case when the derivative $F_u(\omega, \mu, 0)$ has a two-dimensional kernel.

To be precise, suppose F satisfies (1.1) and corresponding to $(\omega_0, \mu_0) \in \mathbb{R}^2$, let us set

$$L = F_u(\omega_0, \mu_0, 0), \ V = \text{Ker}(L) \text{ and } R = R(L).$$

We will say that L (or F) satisfies assumption (II) if

(II-i) $\dim(V) = 2$,
(II-ii) R is closed and $\text{codim}(R) = 2$.

If W, (respectively Z) denotes a complementary subspace of V in X (resp. of R in Y), one has

$$X = V \oplus W, \ Y = Z \oplus R,$$

with $\dim[Z] = 2$. Let P and Q denote the conjugate projections onto

Z and R, respectively. We also set $M = F_{u,\mu}(\omega_0, \mu_0, 0)$ and $N = F_{u,\omega}(\omega_0, \mu_0, 0)$ (See Remark 1.4.4).

Theorem 1.1 (Abstract Hopf Bifurcation Theorem) *Suppose F satisfies (1.1) and let (ω_0, μ_0) be such that $L = F_u(\omega_0, \mu_0, 0)$ satisfies (II). Moreover we assume that there exists $\overline{v} \in V$ such that*

$$PM\overline{v} \text{ and } PN\overline{v} \text{ are linearly independent.} \qquad (1.2)$$

Then (ω_0, μ_0) is a bifurcation point for F.

Remark 1.2 Assumption (1.2) does not depend on the choice of the projector P (on the choice of the complementary subspace Z of R) Indeed, (1.2) is obviously equivalent to require that $M\overline{v}$ and $N\overline{v}$ are linearly independent in Y/R.

Proof of Theorem 1.1. We use the Lyapunov–Schmidt procedure. Let $Q = I - P$ and set $u = v + w$, with $v \in V$ and $w \in W$. The equation $F(\omega, \mu, u) = 0$ splits into the system

$$\left. \begin{array}{l} QF(\omega, \mu, v + w) = 0, \\ PF(\omega, \mu, v + w) = 0. \end{array} \right\} \qquad (1.3)$$

As usual, the Implicit Function Theorem allows us to solve uniquely the first of (1.3) yielding

$$w = \psi(\omega, \mu, v),$$

where ψ is a C^2 function, defined in a neighbourhood of $(\omega_0, \mu_0, 0)$ in $\mathbb{R}^2 \times V$, with values in W, satisfying (see Section 5.3)

$$\psi(\omega, \mu, 0) \equiv 0,$$

$$\psi_v(\omega_0, \mu_0, 0) = 0.$$

Substituting into the second of (1.3), we find

$$PF(\omega, \mu, v + \psi(\omega, \mu, v)) = 0. \qquad (1.4)$$

Taking advantage of (1.2) let us seek solutions of (1.4) in the form $v = s\overline{v}$, with $s \in \mathbb{R}$. So we are led to

$$h(\omega, \mu, s) := PF(\omega, \mu, s\overline{v} + \psi(\omega, \mu, s\overline{v})) = 0.$$

Plainly one has $h(\omega, \mu, 0) = PF(\omega, \mu, \psi(\omega, \mu, 0)) = PF(\omega, \mu, 0) \equiv 0$. Let us set

$$h(\omega, \mu, s) = \chi(\omega, \mu, s)s$$

and note that χ is a function with values in Z, of class C^1 because h is of class C^2 and $h(\omega, \mu, 0) = 0$. With straight forward calculation one has

$$\chi(\omega, \mu, 0) = \frac{\partial h}{\partial s}(\omega, \mu, 0) = PF_u(\omega, \mu, 0)[\overline{v} + \psi_v(\omega, \mu, 0)\overline{v}].$$

Since $\psi_v(\omega_0, \mu_0, 0) = 0$, it follows that

$$\chi(\omega_0, \mu_0, 0) = PF_u(\omega_0, \mu_0, 0)[\bar{v}] = PL\bar{v} = 0.$$

Moreover there results

$$\frac{\partial \chi}{\partial \mu}(\omega_0, \mu_0, 0) = PF_{u,\mu}(\omega_0\mu_0, 0)[\bar{v} + \psi_v(\omega_0, \mu_0, 0)\bar{v}]$$

$$+ PF_u(\omega_0, \mu_0, 0)[\psi_{v,\mu}(\omega_0, \mu_0, 0)\bar{v}].$$

Using again the fact that $\psi_v(\omega_0, \mu_0, 0) = 0$, and since $PF_u(\omega_0, \mu_0, 0)u = PLu = 0$ for all u, we find that

$$\frac{\partial \chi}{\partial \mu}(\omega_0, \mu_0, 0) = PF_{u,\mu}(\omega_0, \mu_0, 0)[\bar{v}] = M\bar{v}.$$

Similarly, one finds

$$\frac{\partial \chi}{\partial \omega}(\omega_0, \mu_0, 0) = PF_{u,\omega}(\omega_0, \mu_0, 0)[\bar{v}] = N\bar{v}.$$

From assumption (1.2) we deduce that the Jacobian $|\partial\chi/\partial(\omega, \mu)|$ evaluated at (ω_0, μ_0) is different from 0. Therefore the Implicit Function Theorem applied to χ allows us to solve (locally) the equation $\chi = 0$ with respect to (ω, μ) in functions of s. More precisely we can find C^1 functions $\omega(s)$ and $\mu(s)$ (defined in a neighborhood of $s = 0$) such that $\omega(0) = \omega_0, \mu(0) = \mu_0$, and $\chi(\omega(s), \mu(s), s) = 0$. It follows that

$$PF(\omega(s), \mu(s), s\bar{v} + \psi(\omega(s), \mu(s), s\bar{v})) = 0$$

and hence the branch $u = u_s = s\bar{v} + \psi(\omega(s), \mu(s), s\bar{v})$, gives rise to a family of nontrivial solutions of $F = 0$. Since $u_s \neq 0$ for $s \neq 0$ and $u_s \to \psi(\omega_0, \mu_0, 0) = 0$ as $s \to 0$, it follows that (ω_0, μ_0) is a bifurcation point.

Remark 1.3 Let us note for future reference that in the cartesian representation of the bifurcating branch, $u_s = s\bar{v} + \psi(\omega(s), \mu(s), s\bar{v})$, the remainder term ψ satisfies $\psi(\omega, \mu, 0) \equiv 0$ and $\psi_v(\omega_0, \mu_0, 0) = 0$.

2 Nonlinear oscillations of autonomous systems

Consider the autonomous system

$$\frac{du}{dt} = f(\mu, x) \tag{S_μ}$$

under the assumption that $f \in C^2(\mathbb{R} \times \mathbb{R}^n, \mathbb{R}^n)$ and

$$f(\mu, 0) \equiv 0 \tag{2.1}$$

As a consequence of (2.1), (S_μ) possesses for any $\mu \in \mathbb{R}$ the *trivial* solution $u(t) \equiv 0$ and we want to find the possible values of μ where

there is a *branching off* of nonconstant periodic solutions of (S_μ). In
order to employ the abstract Hopf Bifurcation Theorem, we introduce,
as anticipated in Section 1, an additional real parameter $\omega > 0$ and
consider the auxilary system

$$\omega \frac{du}{dt} = f(\mu, u). \qquad (S_{\mu,\omega})$$

If u is a 2π-periodic solutions of $(S_{\mu,\omega})$, then $\tilde{u}(t) = u(\omega t)$ is a $(2\pi/\omega)$-
periodic solution of (S_μ).

An application of Theorem 1.1 prompts our discussion of the system
$(S_{\mu,\omega})$.

Let

$$X = \{u \in C^1(\mathbb{R}, \mathbb{R}^n) : u(t + 2\pi) = u(t)\},$$
$$Y = \{y \in C(\mathbb{R}, \mathbb{R}^n) : y(t + 2\pi) = y(t)\}$$

and define $F : \mathbb{R}^2 \times X \to Y$ by setting

$$F(\omega, \mu, u) = \omega u' - f(\mu, u). \qquad (2.2)$$

where, here and hereafter, the prime ′ stands for d/dt. Note that $F \in$
$C^2(\mathbb{R}^2 \times X, Y)$ and $F(\omega, \mu, 0) \equiv 0$ for all $(\omega, \mu) \in \mathbb{R}^2$.

It is perhaps worthwile to make precise the relationship between the
bifurcation of solutions of (2.2) and the branching of periodic solutions
of (S_μ). Let $(\omega_0, \mu_0) \in \mathbb{R}^2$, $\omega_0 > 0$, be a bifurcation point for F,
according to the definition given in the preceding section. Then there
exist sequences $\omega_n \to \omega_0$, $\mu_n \to \mu_0$ and $u_n \in X$ such that $u_n \to 0$, $u_n \neq$
0, satisfying $F(\omega_n, \mu_n, u_n) = 0$. It follows that u_n is a sequence of 2π-
periodic solutions of (S_{μ_n, ω_n}). Moreover setting $\tilde{u}_n(t) = u_n(\omega_n t), \tilde{u}_n$
turns out to be a sequence of periodic solutions of (S_{μ_n}), whose period
$T_n = 2\pi/\omega_n$ tends to $T_0 = 2\pi/\omega_0$. Of course, the *amplitude* of the
orbit $\tilde{u}_n$, namely $\sup_{t\in\mathbb{R}} |\tilde{u}_n(t)| (= \sup_{t\in\mathbb{R}} |u_n(t)|)$, tends to 0 as $s \to 0$.
Hereafter, if the pair (ω_0, μ_0) is a bifurcation point for $F(\omega, \mu, u) = \omega u' -$
$f(\mu, u)$, we will say that *from $(\mu_0, 0)$ there bifurcate periodic solutions of*
(S_μ) *with period close to* $2\pi/\omega_0$.

In order to find bifurcations for F, we have to consider the linearized
system $F_u(\omega, \mu, 0)u = 0$, namely

$$\omega u' - A_\mu u = 0, \quad A_\mu := f_x(\mu, 0).$$

We shall assume there exists μ_0 such that, setting $A_0 = A_{\mu_0}$, we get

(A_0-1) A_0 *is non-singular and has a pair of simple purely imaginary
eigenvalues* $\pm i\omega_0$, $\omega_0 > 0$;

(A_0-2) A_0 *has no other eigenvalues of the form* $\pm ik\omega_0$, $k \in \mathbb{N}$, $k \neq 1$.

Keeping the notation of Section 1, we let $L = F_u(\omega_0, \mu_0, 0)$, $V =$
$\text{Ker}(L)$ and $R = R(L)$, and start by proving a lemma.

Lemma 2.1 *If $(A_0\text{-}1\text{-}2)$ hold, then L satisifies assumption* (II) *of Section 1.*

Proof. We follow a procedure similar to that used in the proof of Lemma 6.3.3. It is convenient to work with complex notation, However, let us point out that all the quantities we will consider turn out to be real.

We begin with $\mathrm{Ker}(L)$. The equation $Lu = 0$, $u \in X$, leads us to find 2π-periodic solutions of

$$\omega_0 \frac{du}{dt} - A_0 u = 0. \tag{2.3}$$

Using the Fourier method, let us put $u = \sum u_k e^{ikt}$, where $u_k \in \mathbf{C}^n$ and $u_{-k} = u_k^*$ (here and below ξ^* denotes the complex conjugate of ξ). Substituting into (2.3) then, formally, we find

$$(ik\omega_0 I - A_0)u_k = 0, \quad k \in \mathbf{Z}. \tag{2.4.k}$$

From $(A_0\text{-}1\text{-}2)$ it follows that the matrix $(ik\omega_0 I - A_0)$ is invertible for any $k \neq \pm 1$ and hence we find $u_k = 0$ for all $k \neq \pm 1$. For $k = \pm 1$ (2.4) gives rise to

$$(\pm i\omega_0 I - A_0)u_{\pm 1} = 0. \tag{2.5}$$

By assumption both $\pm i\omega_0$ are simple eigenvalues of A_0. If $\xi \in \mathbf{C}^n$ is a vector spanning $\mathrm{Ker}(i\omega_0 I - A_0)$ then $\mathrm{Ker}(-i\omega_0 I - A_0)$ is spanned by its conjugate ξ^*. It follows that $\mathrm{Ker}(L)$ is spanned by the functions ξe^{it} and $\xi^* e^{-it}$. Plainly, to obtained real-valued functions we shall take

$$u = a\xi e^{it} + a^* \xi^* e^{-it}, \quad a \in \mathbf{C},$$

and hence the real dimension of V is 2.

Next, to study $R = R(L)$, let us consider the equation $Lu = h$, with $u \in X$ and $h \in Y$. Using the Fourier method again, we are led to the system of infinitely many equations

$$(ik\omega_0 I - A_0)u_k = h_k \quad (k \in \mathbf{Z}), \tag{2.5.k}$$

where h_k denotes the Fourier coefficient of h. As before, $(A_0\text{-}1\text{-}2)$ imply that $(2.5.k)$ is uniquely solvable for any $k \neq \pm 1$. Moreover, since

$$(ik\omega_0 I - A_0)^{-1} = \frac{1}{ik\omega_0} I + O\left(\frac{1}{k^2}\right),$$

the same arguments used in Lemma 6.3.3 show that

$$\sum_{k \neq \pm 1} \eta_k e^{ikt} \in X, \tag{2.6}$$

where $\eta_k = (ik\omega_0 I - A_0)^{-1} h_k$ $(k \neq \pm 1)$.

Let us denote by Π (resp. Π^*) the spectral projectors associated to ξ and ξ^*, respectively. Then the equation

$$(\pm i\omega_0 I - A_0)u_{\pm 1} = h_{\pm 1}$$

has a solution if and only if $\Pi(h_1) = \Pi^*(h_{-1}) = 0$. Using (2.6) also we infer that $y \in W$ if and only if $\Pi(h_1) = \Pi^*(h_{-1}) = 0$. This proves that $\operatorname{codim}(R) = 2$.

It is worth noting explicitly that from the preceding proof it follows that the projection P onto Z is nothing but

$$Pu = \Pi(u_1)e^{it} + \Pi^*(u_{-1})e^{-it} \qquad (2.7)$$

In order to show that F satisfies (1.2) of Section 1, we take $\bar{v} = \xi e^{it} + \xi^* e^{-it}$ and evaluate $PF_{u,\mu}(\omega_0, \mu_0, 0)\bar{v}$ and $PF_{u,\omega}(\omega_0, \mu_0, 0)\bar{v}$. For this, let us start by noticing that

$$F_u(\omega, \mu, 0)\bar{v} = \omega \frac{d\bar{v}}{dt} - A_\mu \bar{v} = i\omega(\xi e^{it} - \xi^* e^{-it}) - A_\mu(\xi e^{it} + \xi^* e^{-it}). \quad (2.8)$$

Taking the derivative with respect to ω one finds

$$PN\bar{v} = PF_{u,\omega}(\omega_0, \mu_0, 0)\bar{v} = i\xi e^{it} - i\xi^* e^{-it}. \qquad (2.9)$$

In order to evaluate $M\bar{v} = F_{u,\mu}(\omega_0, \mu_0, 0)\bar{v}$, some preliminares are in order. Let us denote by $\lambda(\mu) = \alpha(\mu) + i\beta(\mu)$ the branch of eigenvalues of A_μ such that $\lambda(\mu_0) = i\omega_0$ (namely $\alpha(\mu_0) = 0$ and $\beta(\mu_0) = \omega_0$). Since f is C^2 then α and β are C^1 functions of μ. Moreover, taking into account that $i\omega_0$ is simple, it is easy to verify that one can associate to A_μ a family ξ_μ of eigenvectors such that

(i)   the map $\mu \to \xi_\mu$ is C^1,

(ii) $A_\mu[\xi_\mu] = \lambda(\mu)\xi_\mu$,

(iii) $\xi_{\mu_0} = \xi$.

Let $A'_{\mu_0} = (dA_\mu/d\mu)_{\mu=\mu_0}$.

Lemma 2.3   *There result*

$$\left.\begin{array}{l} \Pi A'_{\mu_0}[\xi] = \lambda'(\mu_0)\xi, \\ \Pi^* A'_{\mu_0}[\xi^*] = (\lambda'(\mu_0))^* \xi^*. \end{array}\right\}$$

Proof. We shall prove the former equality. The latter follows similarly.
From (ii) it follows that

$$A_\mu \xi = A_\mu[\xi - \xi_\mu] + A_\mu \xi_\mu = A_\mu[\xi - \xi_\mu] + \lambda(\mu)\xi_\mu.$$

Then, setting $\xi' = (d\xi_\mu/d\mu)_{\mu=\mu_0}$ we get

$$A'_{\mu_0}\xi = A'_{\mu_0}(\xi - \xi_{\mu_0}) - A_0 \xi' + \lambda'(\mu_0)\xi_{\mu_0} + \lambda(\mu_0)\xi'.$$

Recalling that $\xi_{\mu_0} = \xi$ and $\lambda(\mu_0) = i\omega_0$, we infer

$$A'_{\mu_0}\xi = \lambda'(\mu_0)\xi + (i\omega_0 I - A_0)\xi'.$$

If we apply the projector Π, since $\Pi = 0$ on the range of $i\omega_0 I - A_0$, the lemma follows.

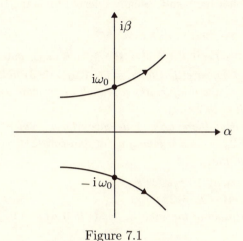

Figure 7.1

The preceding lemma allows us to evaluate $PM\bar{v} = PF_{u,\mu}(\omega_0,\mu_0,0)\bar{v}$. In fact, from (2.8) we infer that

$$M\bar{v} = -A'_{\mu_0}\xi e^{it} - A'_{\mu_0}\xi^* e^{-it}$$

and thus, recalling (2.7),

$$PM\bar{v} = -\Pi A'_{\mu_0}[\xi]e^{it} - \Pi^* A'_{\mu_0}[\xi^*]e^{-it}$$
$$= -(\lambda'(\mu_0)\xi e^{it} + \lambda'^*(\mu_0)\xi^* e^{-it}). \qquad (2.10)$$

Setting $\lambda'(\mu_0) = \alpha'(\mu_0) + i\beta'(\mu_0)$, we find that (2.10) becomes

$$PM\bar{v} = -\alpha'(\mu_0)(\xi e^{it} + \xi^* e^{-it}) - i\beta'(\mu_0)(\xi e^{it} - \xi^* e^{-it}).$$

Then the vectors (2.9) and (2.10) are linearly dependent (over $\mathbb{R}$) if and only if there is a constant $c \in \mathbb{R}$ such that

$$ic = -\alpha'(\mu_0) - i\beta'(\mu_0)$$

which yields $\alpha'(\mu_0) = 0$. In conclusion we can state the following.

Lemma 2.4 *Let us suppose, in addition to* (A$_0$–1–2), *that*

(A$_0$ − 3) $\qquad\qquad\qquad \alpha'(\mu_0) \neq 0.$

Then F satisfies assumption (1.2) *of Section 1.*

Remark 2.5 Condition (A$_0$–3) is a "transversality" condition. From the geometrical point of view, it means that the branch of eigenvalues $\lambda(\mu)$ crosses the imaginary axis transversally for $\mu = \mu_0$ (Figure 7.1).

Lemmas 2.3 and 2.4 allow us to apply Theorem 1.1 to F given by (2.2), yielding the following result, usually referred to as the Hopf Bifurcation theorem [Ho].

Theorem 2.6   *Let $f \in C^2(\mathbb{R} \times \mathbb{R}^n, \mathbb{R}^n)$ be such that $f(\mu, 0) = 0$ for all $\mu \in \mathbb{R}$, and suppose that for $\mu = \mu_0$, $(A_0$–1–2–3$)$ hold.*

Then from μ_0 there bifurcates a branch of periodic solutions of (S_μ), with period close to $2\pi/\omega_0$.

More precisely, there exist a neighbourhood J of $s = 0$, functions $\omega(s)$, $\mu(s) \in C^1(J)$, and a family u_s of non-constant, periodic solutions of $(S_{\mu(s)})$ such that

(i) $\omega(s) \to \omega_0, \mu(s) \to \mu_0$ as $s \to s_0$,
(ii) u_s has period $T_s = 2\pi/\omega(s)$,
(iii)   the amplitude of the orbit u_s tends to 0 as $s \to 0$.

Remark 2.7   One immediately verifies that the same conclusion holds taking $\bar{v} = \xi e^{it(t+\alpha)} + \xi^* e^{-i(t+\alpha)}$ for any $\alpha \in \mathbb{R}$. This fact is related to the fact that (S_μ) being autonomous and therefore invariant under the time translation $\theta \to u(. + \theta)$, the bifurcation set is also.

Example 2.8   Let us consider the Van der Pol equation

$$\frac{d^2 x}{dt^2} - (\mu - 3x^2)\frac{dx}{dt} + x = 0, \tag{2.11}$$

where μ is a real parameter. Equation (2.11) is equivalent to the first-order system

$$\left. \begin{array}{l} \dfrac{dx}{dt} = y + \mu x - x^3, \\[2mm] \dfrac{dy}{dt} = -x. \end{array} \right\} \tag{2.12}$$

The linearized system is

$$\left. \begin{array}{l} \dfrac{dx}{dt} = y + \mu x, \\[2mm] \dfrac{dy}{dt} = -x. \end{array} \right\} \tag{2.13}$$

In this case one has $A_\mu = \begin{bmatrix} \mu & 1 \\ -1 & 0 \end{bmatrix}$ whose eigenvalues are $\lambda(\mu) = \frac{1}{2}[\mu \pm \sqrt{(\mu^2 - 4)}]$. For $\mu = 0$ there results $\lambda(0) = \pm i$. Plainly, $(A_0$–1–2–3$)$ hold true and Theorem 2.6 applies with $\mu_0 = 0$ and $\omega_0 = 1$.

We note that, in this specific case, taking advantage of dealing with a system in $\mathbb{R}^2$, it is possible to perform a direct analysis of the characteristics of (2.12) in the "phase" plane (x, y). One finds easily that

(1)   for any $\mu \leq 0$ (2.12) has no periodic trajectories but the trivial one $x = 0$, $y = 0$,

(2)   for any $\mu > 0$ (2.12) has a unique periodic solution, which is asymptotically stable. See Figures 7.2 and 7.3.

Note also that the trivial solution $x = 0, y = 0$ is stable for all $\mu < 0$ and unstable for $\mu > 0$.

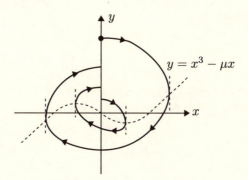

Figure 7.2 Phase portrait of (2.12)

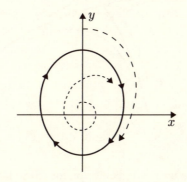

Figure 7.3 The closed orbit of (2.12)

3 The Lyapunov Centre Theorem

Consider the first-order system

$$\frac{du}{dt} = f(u), \qquad \text{(S)}$$

where $f \in C^2(\mathbb{R}^n, \mathbb{R}^n)$. A *singular point* of (S) is a $p \in \mathbb{R}^n$ such that $f(p) = 0$. In this section we will be concerned with the existence of small oscillations of (S) near an equilibrium p, namely periodic solutions of (S) with orbits confined near p. Let us suppose that $p = 0$ is a singular point of (S) and let $A = f'(0)$. If no point of the spectrum of A belongs to the imaginary axis, then the behaviour of the solutions of (S) near $p = 0$ is completely understood. It is possible to show (see, for example [**P**]) that there are two invariant manifolds M and N, with $\dim(M) + \dim(N) = n$ and $M \cap N = \{0\}$ such that for all $q \in M$ (resp. N) the solution of the Cauchy problem

$$\left. \begin{array}{l} \dfrac{du}{dt} = f(u), \\[2mm] u(0) = q, \end{array} \right\}$$

tends to 0 as $t \to +\infty$ (as $t \to -\infty$, respectively). The behaviour of the solutions near $p = 0$ is represented in Figure 7.4.

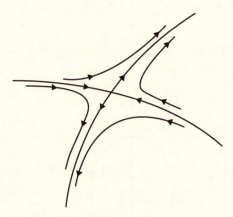

Figure 7.4

As a consequence, a necessary condition for (S) to have closed orbits near a singular point, say $p = 0$, is that A have a pair of purely imaginary eigenvalues $\pm i\omega_0$. However this condition is not sufficient, in general. For example, consider the two-dimensional system

$$\left. \begin{array}{l} x' = -y - x(x^2 + y^2), \\[2mm] y' = x - y(x^2 + y^2). \end{array} \right\} \qquad (3.1)$$

The eigenvalues of A are $\pm i$; on the other hand, if $(x(t), y(t))$ is a solution of (3.1), one has

$$\frac{d}{dt} \left[\frac{1}{2}(x^2 + y^2) \right] = xx' + yy' = -(x^2 + y^2)^2.$$

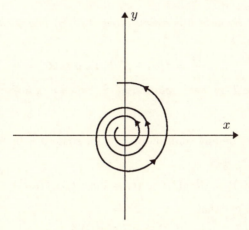

Figure 7.5

It follows that

$$x^2 + y^2 = \frac{1}{2t + c}$$

and hence (3.1) has no periodic solutions (see Figure 7.5).

Lyapunov has shown in a celebrated theorem [**Ly**] that (S) does possess periodic solutions near 0 provided it has a "non-singular" first integral b. Recall that a first integral of (S) is a non-constant real-valued function $b \in C^1(\mathbb{R}^n, \mathbb{R})$ such that $b(u(t)) \equiv$ constant for any solution $u(t)$ of (S). We point out that, in view of the uniqueness of the solutions of the Cauchy problem

$$\frac{du}{dt} = f(u), \quad u(0) = p,$$

b is a first integral of (S) if and only if

$$f(p) \cdot \nabla b(p) = 0, \quad \text{for all } p \in \mathbb{R}^n. \tag{3.2}$$

Examples of such systems are the second-order *conservative* (or *gradient*) systems, namely systems like

$$\frac{d^2 u}{dt^2} + \nabla U(u) = 0, \tag{3.3}$$

or the *Hamiltonian systems*

$$\left.\begin{aligned} x' &= -H_y(x, y), \\ y' &= H_x(x, y), \end{aligned}\right\} \tag{HS}$$

where $(x, y) \in \mathbb{R}^n \times \mathbb{R}^n$. In fact, the Hamiltonian itself $H = H(x, y)$ is a first integral of (HS). Note that (3.3) is a particular case of (HS): it suffices to take $H(x, y) = \frac{1}{2}|y|^2 + U(x)$.

The following lemma shows the role played by the first integral.

Lemma 3.1 *Suppose b is a first integral of* (S) *and consider the modified system*

$$\frac{du}{dt} = f(u) + \mu \nabla b(u), \ \mu \in \mathbb{R}. \tag{$(S)_\mu$}$$

If $u = u(t)$ is a T-periodic solution of (S_μ) then u is in fact a T-periodic solution of (S).

Proof. Let $u(t)$ be any solution of (S_μ) for some $\mu \neq 0$. Setting $\beta(t) = b(u(t))$ one has

$$\beta'(t) = \frac{d}{dt} b(u(t)) = \nabla b(u(t)) \cdot u'(t) = \nabla b(u(t)) \cdot f(u(t)) + \mu |\nabla b(u(t))|^2$$

Using (3.2) we get that

$$\beta'(t) = \mu |\nabla b(u(t))|^2.$$

If, for example, $\mu > 0$, then $\beta(t)$ is non-decreasing. In the other hand, since x is T-periodic, we deduce $\beta(0) = b(u(0)) = b(u(T)) = \beta(T)$. Hence

$$\beta'(t) = \mu |\nabla b(u(t))|^2 \equiv 0$$

and u solves (S).

Lemma 3.1 suggests we seek small oscillations of a system with a first integral as periodic solutions of (S_μ) above, bifurcating from $\mu_0 = (0,0)$. The following theorem gives conditions under which such a bifurcation occurs.

Theorem 3.2 (Lyapunov Centre Theorem) *Suppose that $f \in C^2(\mathbb{R}^n, \mathbb{R}^n)$ is such that $f(0) = 0$. Letting $A = f'(0)$, we suppose that*

(A-1) *A is nonsingular and has a pair of simple eigenvalues $\pm i\omega_0$;*

(A-2) *for all $k \in \mathbb{Z}$, $k \neq \pm 1$, $ik\omega_0$ is not an eigenvalue of A.*

Moreover, let us assume that (S) *has a first integral $b \in C^2(\mathbb{R}^n, \mathbb{R})$ such that $b''(0)$ is non-singular.*

Then (S) *possesses small oscillations near $p = 0$.*

More precisely, there exist a neighbourhood J of $s = 0$, a function $\omega(s) \in C^1(J)$, and a family u_s of non-constant, periodic solutions of (S) *such that*

(i)   *$\omega(s) \to \omega_0$, as $s \to 0$;*

(ii) *u_s has period $T_s = 2\pi/\omega(s)$;*

(iii) *the amplitude of the orbit u_s tends to 0 as $s \to 0$.*

Proof. According to Lemma 3.1, we can replace (S) with (S_μ) which can be studied by means of Theorem 2.6, with $f(\mu, u) = f(u) + \mu \nabla b(u)$. First of all we note that from (3.2), and recalling that b is C^2 here, it follows that

$$f'(\xi)y \cdot \nabla b(\xi) + f(\xi) \cdot \nabla b''(\xi)y = 0, \quad \text{for all } \xi, y \in \mathbb{R}^n. \tag{3.4}$$

Putting $\xi = 0$, one has

$$Ay \cdot \nabla b(0) + f(0) \cdot b''(0)y = 0, \quad \text{for all } y \in \mathbb{R}^n.$$

Since $f(0) = 0$ and A is non-singular, it follows that $\nabla b(0) = 0$. As a consequence, we infer that $f(\mu, 0) = f(0) + \mu \nabla b(0) = 0$. Moreover, setting $B = b''(0)$, one has

$$A_\mu = f_x(\mu, 0) = A + \mu B.$$

We shall apply Theorem 2.6 with $\mu_0 = 0$. Since $A_0 = A_{\mu_0} = A$, (A$_0$–1–2) follow from (A–1–2). It remains to verify that (A$_0$-3) holds.

First, let us consider (3.4). Since f is continuously differentiable, $f(0) = 0$ and b'' is continuous, then it is easy to verify that the map $\xi \to f(\xi) \cdot b''(\xi)y$ is differentiable at $\xi = 0$ with derivative $f'(0)[.] \cdot b''(0)y = A[.] \cdot By$. Hence we can differentiate (3.4) at $\xi = 0$, yielding

$$f''(0)[y, z] \cdot \nabla b(0) + Ay \cdot Bz + Az \cdot By = 0 \text{ for all } y, z \in \mathbb{R}^n.$$

Since $\nabla b(0) = 0$ it follows that

$$Ay \cdot Bz + Az \cdot By = 0 \text{ for all } y, z \in \mathbb{R}^n,$$

that is (note that B is symmetric),

$$A^T B + BA = 0. \tag{3.5}$$

Then, up to a change of coordinates, the matrix A has the form

$$A = \begin{bmatrix} S & 0 \\ 0 & R \end{bmatrix}$$

with

$$S = \begin{bmatrix} 0 & -\omega_0 \\ \omega_0 & 0 \end{bmatrix}$$

and R does not contain $\pm i\omega_0$ in its spectrum, because $\pm i\omega_0$ are simple eigenvalues of A. Let us write

$$B = \begin{bmatrix} U & M \\ M^T & C \end{bmatrix}$$

where U (resp. C) is a symmetric 2×2 (resp. $(n-2) \times (n-2)$) matrix. From (3.5) it follows readily that

$$SU = US \tag{3.6'}$$

and

$$SM = MR. \tag{3.6''}$$

Recalling that $\omega_0 \neq 0$, from (3.6') one deduces with elementary calculations that there is $\delta \in \mathbb{R}$ such that

$$U = \begin{bmatrix} \delta & 0 \\ 0 & \delta \end{bmatrix}.$$

From (3.6'') and using the fact that $\omega_0 \neq 0$ and $\pm i\omega_0$ are not eigenvalues of R, one infers that the $2 \times (n-2)$ matrix M is the 0 matrix. [†]
   To see this, let $X, Y \in \mathbb{R}^{n-2}$ denote the two rows of M.

$$M = \begin{bmatrix} X \\ Y \end{bmatrix}$$

Then from (3.6'') it follows X, Y satisfy the system:

$$\begin{cases} XR + \omega_0 Y = 0 \\ YR - \omega_0 X = 0 \end{cases}$$

One finds $X = \omega_0^{-1} Y R$ and hence $Y(R^2 + \omega_0^2 I) = 0$. Since $\pm i\omega_0$ are not eigenvalues of R, then one infers that $Y = X = 0$.
   From the preceding arguments we deduce that B has, with respect to the same basis used for (3.5), the form

$$B = \begin{bmatrix} \delta & 0 & \\ 0 & \delta & 0 \\ & 0 & C \end{bmatrix}$$

where $\delta \neq 0$, because B is non-singular.
   Consequently there results

$$A + \mu B = \begin{bmatrix} \mu\delta & -\omega_0 & \\ -\omega_0 & \mu\delta & 0 \\ & 0 & R + \mu C \end{bmatrix},$$

and hence the branch of eigenvalues $\lambda(\mu)$ such that $\lambda(0) = i\omega_0$ is given by

$$\lambda(\mu) = \mu\delta + i\omega_0.$$

This proves that $(A_0\text{-}3)$ holds true. An application of Theorem 2.6, jointly with Lemma 3.1, yields the existence of a C^1 function $\omega(s) \to \omega_0$ and of family u_s of non-constant solutions of (S) with period $T_s = 2\pi/\omega(s)$, such that the amplitude of u_s tends to 0 as $s \to 0$.

Remarks 3.3
   (i) The above result being local in nature, it would be sufficient to consider in Theorem 3.2 a vector field f and a first integral b defined in a neighbourhood of 0 in $\mathbb{R}^n$.
   (ii) If A has several purely imaginary eigenvalues $\pm i\omega_k, \omega_k > 0, k =$

[†] More generally, it is possible to show that *if R and S are square matrices having disjoint spectra and if M is a matrix such that SM = MR, then M = 0.*

$1, \ldots, m$, the non-resonance condition (A2) is always satisfied at $\pm i\omega^*$, where $\omega^* = \max\{\omega_k, 1 \leq k \leq n\}$.

(iii) It has been proved by J. Moser [**Mo**] that non-resonance conditions (A1–2) can be eliminated at the expense of the existence of a first integral $b \in C^2(\mathbb{R}^n, \mathbb{R})$ such that $b''(0)$ *is positive-definite*. The following example (see [**Mo**]; see also [**MW**]) shows that, in this more general form, if $b''(0)$ is merely nondegenerate, (S) may have no periodic solutions at all. Let $x, y \in \mathbb{R}^2$, $x = (x_1, x_2)$, $y = (y_1, y_2)$ and consider the Hamiltonian system (HS) with Hamiltonian

$$H(x, y) = \frac{1}{2}(x_1^2 - x_2^2 + y_1^2 - y_2^2) + (|x|^2 + |y|^2)\mathcal{B}(x, y),$$

where

$$\mathcal{B}(x, y) = (y_1 y_2 - x_1 x_2).$$

Here the matrix A has the form

$$A = \begin{bmatrix} 0 & 0 & -1 & 0 \\ 0 & 0 & 0 & 1 \\ 1 & 0 & 0 & 0 \\ 0 & -1 & 0 & 0 \end{bmatrix}$$

and has double eigenvalues $\pm i$. If $x = x(t)$ and $y = y(t)$ is a solution of (HS), there results

$$\frac{\mathrm{d}}{\mathrm{d}t}(x_1 y_2 + y_1 x_2) = -4[\mathcal{B}(x, y)]^2 - (|x|^2 + |y|^2)^2,$$

and therefore (HS) has the trivial solution $x \equiv 0$, $y \equiv 0$ only.

(iv) Since b'' is non-singular, the arguments of Lemma 3.1 show that here the auxiliary parameter $\mu = 0$.

(v) According to Remark 1.3, the family of periodic solutions u_s has the property that

$$\frac{u_s}{s} \to \xi e^{i\omega t} + \xi^* e^{-i\omega t} \quad \text{as } s \to 0$$

where ξ is such that $A\xi = i\omega_0 \xi$.

The Lyapunov Centre Theorem applies both to second-order gradient systems like (3.3) and to Hamiltonian Systems (HS). Let us state explicitly this kind of result.

We consider (HS) with $H \in C^2(\mathbb{R}^n \times \mathbb{R}^n, \mathbb{R})$. Set $z = (x, y) \in \mathbb{R}^{2n}$, $H(z) = H(x, y)$ and $\nabla H(z) = (H_x(z), H_y(z))$. If J denotes the symplectic matrix (i.e. $J : (x, y) \to (-y, x)$), then (HS) can be written in the more compact form

$$\frac{\mathrm{d}z}{\mathrm{d}t} = J\nabla H(z).$$

Since in (HS) the Hamiltonian H is a constant of the motion, namely

$H(z(t)) \equiv$const. for all solutions of (HS), it makes sense to look for periodic solutions of (HS) on the Hamiltonian surface $H(z) = h$.

We suppose that

(H0) $H(0) = 0$, $\nabla H(0) = 0$ and $H''(0) > 0$ (that is $H''(0)$ is positive-definite),

(H1) $JH''(0)$ has n pairs of purely imaginary simple eigenvalues

$$\pm i\omega_k, \ k = 1, 2, \dots, n,$$

such that ω_i/ω_j is not an integer for all $i \neq j$.

Theorem 3.4 *Suppose that* $H \in C^2(\mathbb{R}^n \times \mathbb{R}^n, \mathbb{R})$ *satisfies* (H0–1). *Then for all* $\varepsilon > 0$ *small enough* (HS) *has* n *(geometrically) distinct periodic orbits on the surface* $H(z) = \varepsilon$. *More precisely, the surface* $H(z) = \varepsilon$ *carries* n *distinct periodic orbits* z_k *whose periods tend to* $2\pi/\omega_k$, $k = 1, 2, \dots, n$.

Proof. For all $k = 1, 2, \dots, n$, we can apply Theorem 3.2 with $f = J\nabla H$, $A = JH''(0)$ and $b = H$. Indeed, (H0) implies, in particular, that $b''(0) = H''(0)$ is non-singular; and (H1) implies that (Ai–ii) hold true for all $k = 1, 2, \dots, n$.

Then there exist n branches $z_{k,s}$, $k = 1, 2, \dots, n$. of periodic solutions of (HS) with period $T_{k,s}$ converging to $2\pi/\omega_k$ as $s \to 0$; moreover,

$$\|z_{k,s}\|_{L^\infty} \to 0 \quad \text{as } s \to 0. \tag{3.7}$$

In addition, $z_{k,s}$ depends in a C^1 fashion on s and (see Remark 3.3(v))

$$\lim_{s \to 0} \frac{z_{k,s}}{s} = v_k := \xi_k e^{i\omega t} + \xi_k^* e^{-i\omega t}, \quad k = 1, 2, \dots, n \tag{3.8}$$

where $A\xi_k = i\omega_k \xi_k$.

Since H is a first integral of (HS), then $H(z_{k,s}(t))$ is independent of t. We set

$$h_k(s) = H(z_{k,s}(0)).$$

From (3.7) one immediately deduces that $h_k(s) \to H(0) = 0$.

Moreover, since $z_{k,s}$ is C^1 with respect to s and $H'(0) = 0$, it follows readily that h_k is twice differentiable at $s = 0$ and, using also (3.8), one finds

$$h_k''(0) = H''(0)w_k \cdot w_k$$

where $w_k = \xi_k + \xi_k^*$.

Since $H''(0) > 0$, it follows that for all $\varepsilon > 0$ small enough and any $k = 1, 2, \dots, n$, the equation $h_k(s) = \varepsilon$ has a solution $s = s(k, \varepsilon)$ and

$$s(k, \varepsilon) \to 0 \text{ as } \varepsilon \to 0 \quad k = 1, 2, \dots, n.$$

Correspondingly we find n solutions $z_{k,\varepsilon} = z_{k,s(k,\varepsilon)}$ $(k = 1, 2, \ldots, n)$ of (HS) such that $H(z_{k,\varepsilon}) = \varepsilon$. Finally, from (3.8) we also deduce that, for ε small, the orbit of $z_{k,\varepsilon}$ is close to that of $s\tilde{v}_k$, that is to that of sv_k, up to higher-order terms; then the $z_{k,\varepsilon}$ $(k = 1, 2, \ldots, n)$ correspond to *geometrically distinct orbits*. This completes the proof of the theorem.

Remarks 3.5

(i) As in Remark 3.3 (i), H could be defined in a neighbourhood of 0 in $\mathbb{R}^n$.

(ii) Theorem 3.4 has been extended by Weinstein [**W**] (see also [**Mo**]) who proved the following result. *Suppose H satisfies* (H0). *Then for all $\varepsilon > 0$ small enough* (HS) *has n distinct periodic orbits on the surface $H(z) = \varepsilon$.* In comparison with the result of Moser recalled in Remark 3.3 (iii), one has to point out that in the case of a general conservative system one can exibit examples where (S) has only one solution on each surface $b = \varepsilon$.

4 The restricted three-body problem

One of the most classical application of the Lyapunov Centre Theorem is to the existence of small oscillations near the equilibrium points of the planar restricted three-body problem. This problem deals with three-bodies P_1, P_2 (called *primaries*) and Q, with masses M_1, M_2 and M_3, respectively, under the action of the Newton Gravitational Law. To make the problem more feasible, one considers the *restricted problem*, which is concerned with the case when the mass of one particle is negligible with respect to the others. If, say, $M_3 = 0$ then P_1 and P_2 are not influenced by Q and they move according to the solutions of a two-body problem. Our aim is to study the motion of Q under the attraction of the two primaries.

Actually, we shall make some further simplifications. First of all, we suppose that the primaries move on circles, rather than more general elliptical orbits, with constant angular velocity γ. Moreover we will assume that the motion of Q occurs on the same plane as that of P_1, P_2. This problem is usually called the *restricted planar three-body problem*. Even with these simplifications, it is still quite interesting, because many problems arising in celestial mechanics fit in this frame.

Let us introduce a rotating coordinate system xOy (Fig. 7.6), such that (i) the origin O coincides with the barcentre of P_1 and P_2 and (ii) P_1 and P_2 are at rest on the x-axis. With a suitable choice of the units,

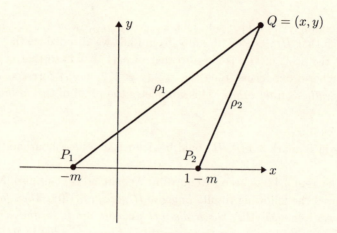

Figure 7.6

we can take $M_1 + M_2 = 1$, $\gamma = 1$ and g (the gravity constant) $= 1$. We also set $M_2 = m$ in such a way that $P_1 = (-m, 0)$ and $P_2 = (1 - m, 0)$ and let (x, y) denote the coordinates of Q and

$$\rho_1 = \sqrt{[(x + m)^2 + y^2]},$$
$$\rho_2 = \sqrt{[(x + m - 1)^2 + y^2]}$$

the distances from Q to P_1 and P_2, respectively.

The third body Q is subjected to combined action of the centrifugal and Coriolis forces and to those due to the Newtonian attraction, corresponding to the potential

$$U(x, y) = \frac{1 - m}{\rho_1} + \frac{m}{\rho_2}.$$

In conclusion, we find the system

$$\left.\begin{array}{l} x'' - 2y' - x = U_x(x, y), \\ y'' + 2x' - y = U_y(x, y), \end{array}\right\} \tag{4.1}$$

where, here and hereafter, primes $\prime$ denote d/dt.

Equilibrium points

The possible equilibria of (4.1) can be found by solving the system

$$\left.\begin{array}{l} -x = U_x(x, y), \\ -y = U_y(x, y), \end{array}\right\}$$

namely the pair of equations

$$-x = -\frac{(x + m)(1 - m)}{\rho_1^3} - \frac{m(x + m - 1)}{\rho_2^3}, \tag{4.2}$$

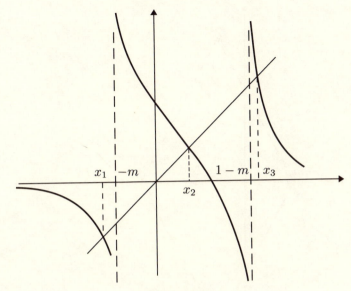

Figure 7.7

$$-y = -\frac{y(1-m)}{\rho_1^3} - \frac{my}{\rho_2^3}. \tag{4.3}$$

The latter is satisfied for $y = 0$. Substituting $y = 0$ into (4.2) we find

$$x = \frac{(x+m)(1-m)}{|x+m|^3} + \frac{m(x+m-1)}{|x+m-1|^3}. \tag{4.4}$$

Equation (4.4) has three solutions, corresponding to the so called *Euler points* L_1, L_2, L_3. (Figure 7.7).

It is also convenient to introduce the potential

$$\Phi(x, y) = \frac{1}{2}(x^2 + y^2) + U(x, y). \tag{4.5}$$

Equations (4.2) and (4.3) are nothing but $\Phi_x = 0$ and $\Phi_y = 0$, respectively. Hence the Euler points are the solutions of $\Phi_x(x, 0) = 0$. One checks immediately that $\Phi_{xx}(L_i) > 0$, $i = 1, 2, 3$. As for Φ_{yy} one finds

$$\Phi_{yy}(x, 0) = 1 - \frac{m}{|x - m + 1|^3} - \frac{1-m}{|x - m|^3}.$$

Since in L_2 both $|x-m|$ and $|x-m+1|$ are < 1 we infer that $\Phi_{yy}(L_2) < 0$. In L_1 and L_3 one finds, with elementary calculations, that $x\Phi_{yy}(x, 0)$ is < 0 in L_3 and > 0 in L_1. In both cases it follows that $\Phi_{yy} < 0$. In other words, letting for $1 \leq j \leq 3$

$$a_j = \Phi_{xx}(L_j), \ b_j = \Phi_{xy}(L_j), \ c_j = \Phi_{yy}(L_j),$$

we get

$$b_j = 0, \ a_j > 0 \text{ and } c_j < 0. \tag{4.6}$$

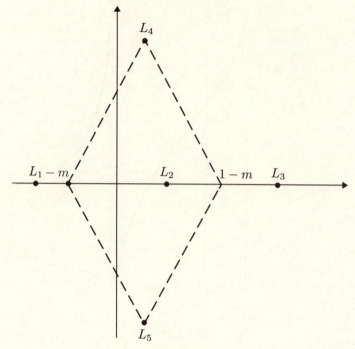

Figure 7.8

Let us come back to (4.2)–(4.3) and look for solutions with $y \neq 0$. Setting $h = 1/\rho_1^3$ and $k = 1/\rho_2^3$ we find

$$x = h(x + m)(1 - m) + km(x + m - 1), \qquad (4.7)$$

$$1 = h(1 - m) + km. \qquad (4.7')$$

Multiplying (4.7') by x and subtracting from (4.7) one has readily $h = k = 1$. Thus there are two more equilibria L_4 and L_5, the *Lagrangian points*, such that P_1, P_2 and L_4 (or L_5) are the vertices of an equilateral triangle (Figure 7.8).

Setting $a = a_{4,5} = \Phi_{xx}(L_{4,5})$, $b = b_{4,5} = \Phi_{x,y}(L_{4,5})$ and $c = c_{4,5} = \Phi_{yy}(L_{4,5})$, one finds readily

$$a = \frac{3}{4}, b = \frac{3\sqrt{2}}{4}(2m - 1), c = \frac{9}{4}. \qquad (4.8)$$

The configuration consisting of the two primaries and a Lagrange point is, for example, that of the system Sun-Jupiter-Trojans (the last are a group of asteroids).

Small oscillations

Let us refer to the system (4.1) which will be written in the form

$$\left.\begin{array}{l} x'' - 2y' = \Phi_x(x,y), \\ y'' + 2x' = \Phi_y(x,y), \end{array}\right\} \tag{4.1'}$$

where Φ is given by (4.5). In order to apply the Lyapunov Centre Theorem, (4.1') has to be transformed into a first-order system. If we set $p = x'$ and $q = y'$, (4.1') becomes

$$\left.\begin{array}{l} x' = p, \\ y' = q, \\ p' = 2q + \Phi_x, \\ q' = -2p + \Phi_y, \end{array}\right\} \tag{4.1''}$$

which is of the form $u' = f(u)$, where $u = (x,y,p,q) \in \mathbb{R}^4$ and f has components

$$f(x,y,p,q) = (p, q, 2q + \Phi_x, -2p + \Phi_y).$$

In terms of the new coordinates, the equilibria are given by

$$u_j = (x_j, y_j, 0, 0), \ 1 \le j \le 5, \ \text{where } (x_j, y_j) = L_j.$$

It is immediately verifiable the system (4.1'') has a first integral (the *Jacobi integral*) given by

$$J(x,y,p,q) = \frac{1}{2}(p^2 + q^2) - \Phi(x,y).$$

The Hessian $J''(u_j)$ is given by (we keep the notation introduced before)

$$J''(u_j) = \begin{bmatrix} -a_j & -b_j & 0 & 0 \\ -b_j & -c_j & 0 & 0 \\ 0 & 0 & 1 & 0 \\ 0 & 0 & 0 & 1 \end{bmatrix}.$$

Consequently

$$\det[J''(u_j)] = \begin{vmatrix} -a_j & -b_j \\ -b_j & -c_j \end{vmatrix} = a_j c_j - b_j^2.$$

Taking into account (4.6) and (4.9) we find

$$\det[J''(u_j)] < 0, \text{ for } j = 1, 2, 3,$$

$$\det[J''(u_j)] = \frac{27}{4}m(1-m) > 0, \text{ for } j = 4, 5,$$

and in any case J'' is nonsingular at each equilibrium point.

It remains to evaluate the matrix $A_j = f'(x_j, y_j, 0, 0)$. We obtain

$$A_j = \begin{bmatrix} 0 & 0 & 1 & 0 \\ 0 & 0 & 0 & 1 \\ a_j & b_j & 0 & 2 \\ b_j & c_j & -2 & 0 \end{bmatrix}.$$

The eigenvalues λ of A_j satisfy the equation

$$\lambda^4 - (a_j + c_j - 4)\lambda^2 + D_j = 0, \tag{4.9}$$

with

$$D_j = \det \begin{pmatrix} -a_j & -b_j \\ -b_j & -c_j \end{pmatrix}.$$

On the Euler points L_j, $j = 1, 2, 3$, we have $b_j = 0$, and (4.9) becomes

$$\lambda^4 - (a_j + c_j - 4)\lambda^2 + a_j c_j = 0. \tag{4.9'}$$

Since in addition $a_j c_j < 0$, then (4.9′) has a unique pair of purely imaginary roots $\lambda = \pm i\omega_j$, $j = 1, 2, 3$, and the Lyapunov Centre Theorem applies without any further restriction yielding the following.

Theorem 4.1 *In a neighbourhood of the Euler points L_j, $j = 1, 2, 3$, the restricted planar three-body problem has a family of periodic solutions whose periods tend to $2\pi/\omega_j$.*

As for the Lagrangian point L_4 (the same holds for the symmetric one L_5) we have (see (4.8))

$$a = \frac{3}{4}, \quad b = \frac{3\sqrt{2}}{4}(2m - 1), \quad c = \frac{9}{4}.$$

The (4.9) becomes

$$\lambda^4 + \lambda^2 + \frac{27}{4}m(1 - m) = 0,$$

which possesses two pairs of imaginary roots $\pm i\omega'$, $\pm i\omega''$, with, say, $0 < \omega' < \omega''$, provided $\frac{27}{4}m(1 - m) < \frac{1}{4}$, namely for all $0 < m < m_0$ (or $1 - m_0 < m < 1$) where $m_0 \approx 0.0385\ldots$ is the smallest roots of $27m(1 - m) = 1$. Let us consider the range $0 < m < m_0$ (for example, in the case when the primaries are Sun and Jupiter, the mass ratio is $m \approx 1/1000$, a value which is widely in the range $(0, m_0 \approx 0.0385)$; the same for the Earth–Moon system where $m \approx 1/82 \approx 0.012$).

Taking $\omega_0 = \omega''$ we can apply the Lyapunov Centre Theorem directly, while when we consider the pulsation ω' we have to require the *non-resonance condition* $\omega''/\omega' \notin \mathbb{N}$, namely that $\omega'' \neq k\omega'$ for all $k \in \mathbb{N}$. This leads to excluding the solutions of

$$\left.\begin{aligned} \omega''^2 &= k^2\omega'^2, \\ \omega''^2 + \omega'^2 &= 1, \\ \omega''^2\omega'^2 &= \frac{27}{4}m(1 - m). \end{aligned}\right\}$$

This system has solutions whenever m satisfies

$$\frac{27}{4}m(1 - m) = \frac{k^2}{(1 + k^2)^2}. \tag{4.10}$$

In conclusion, if m_k denotes the sequence of solutions of (4.10) such that $m_k \to 0$, we can still apply, for $m \neq m_k$, the Lyapunov Centre Theorem yielding the following.

Theorem 4.2 *Suppose that* $0 < m < m_0$; *then in a neighbourhood of the Lagrange points* $L_{4,5}$ *the restricted planar three-body problem has a family of periodic solutions whose periods tend to* $2\pi/\omega''$; *if further,* $m \neq m_k$, *then there exists a second family of periodic solutions whose periods tend to* $2\pi/\omega'$.

Note that these periodic solutions correspond to *bounded* trajectories in the inertial frame of reference. For other results on the restricted three-body problem, see for example [SiM].

Remark 4.3 The stability of the linearized system $u' = A_j u$, namely of

$$\left. \begin{array}{l} x'' - 2y' = a_j x + b_j y, \\ y'' + 2x' = b_j x + c_j y, \end{array} \right\} \tag{4.11}$$

at the equilibria $L_j, j = 1, 2, 3, 4, 5$, can be easily discussed.

At the Euler points L_1, L_2, L_3 the matrix A_j has a real positive and a real negative eigenvalue. Thus $L_j (j = 1, 2, 3)$ are unstable equilibria for (4.11) and are said to be *linearly unstable*.

Unlike the preceding case, the matrices A_3, A_4 have, for $0 < m < m_0$, two pairs of purely imaginary eigenvalues. Therefore, at L_4, L_5, (4.11) has bounded orbits only and the Lagrangian points are said to be *linearly stable*.

The question of the (nonlinear) stability of L_4, L_5 is much more delicate. It has been shown there are three exceptional values $m_i (i = 1, 2, 3), 0 < m_1 < m_2 < m_3 < m_0$, such that for all $m \in (0, m_0), m \neq m_i$ $(i = 1, 2, 3)$, the Lagrangian points are stable in the sense of Lyapunov. For more details, see [Mo1].

Problems

1. Let $\mathrm{Inv}(X,Y)$ denote the set of all $A \in L(X,Y)$ that are invertible with inverses $A^{-1} \in L(Y,X)$. Show that

(i) $\mathrm{Inv}(X,Y)$ is open,
(ii) the map $(\cdot)^{-1} : \mathrm{Inv}(X,Y) \to \mathrm{Inv}(Y,X)$ defined by $(\cdot)^{-1}(A) = A^{-1}$ is differentiable and
$$\mathrm{d}(\cdot)^{-1}(A) : H \to -A^{-1}HA^{-1}.$$

2. If f satisfies the Carathéodory condition (C) and
$$|f(x,s)| \leq a(x) + b|s|^\sigma$$
with $\sigma = p/q$, $a \in L^q$ and $b > 0$, show that f is continuous from L^p to L^q.

3. Let $K : \Omega \times \Omega \to \mathbb{R}$ be such that $K(x,y) = K(y,x)$ for all $(x,y) \in \Omega \times \Omega$ and
$$\int_{\Omega \times \Omega} |K(x,y)|^p \mathrm{d}x\mathrm{d}y < \infty.$$
If f satisfies (C) and (2.1) show that the Hammerstein operator
$$H : u(x) \to \int_\Omega K(x,y)f(y,u(y))\mathrm{d}y$$
is continuous from L^p into itself.

4. If, in addition to the conditions of problem 3, f has partial derivative f_s satisfying (C) and (2.7) of Section 1.2, with $p > 2$, show that H is F-differentiable on L^p and
$$\mathrm{d}H(u)[v] = \int_\Omega K(x,y)f_s(y,u(y))v(y)\mathrm{d}y.$$

5. Consider the operator N defined in Section 1.2, eqn (2.19) and assume f satisfies (2.17) therein. Show that if $\sigma < (n+2)/(n-2)$ then N is compact.

Hint Use the compactness of the Sobolev embedding of $H_0^1(\Omega)$ into $L^p(\Omega)$ for any $p < 2n/(n-2)$.

6. Consider $F \in C^1(X,Y)$ with $F(0) = 0$. Let us set $N = \text{Ker}(F'(0))$ and suppose that (i) N has a complementary subspace Z in X, and (ii) $F'(0)$ is onto Y. Show that there are $\varepsilon > 0$, neighbourhoods Θ (resp. V, W) of 0 in N (resp. in Y, Z) and a map $\Phi \in C^1(\Theta \times V, W)$ such that $F(\xi + \Phi(\xi, v)) = v$, for all $(\xi, v) \in \Theta \times V$.

Apply this result to the case in which $X = \mathbb{R}^m, Y = \mathbb{R}^n (m > n)$ and the rank of the matrix $A = F'(0)$ is n, and show that the equation $F(x) = 0$ has a solution $x = \xi + z$, with $z = g(\xi) := \varphi(\xi, 0)$.

7. Consider the boundary-value problem

(P1)
$$\begin{cases} -\Delta u = f(u) \text{ in } \Omega, \\ u = \varepsilon \text{ on } \partial\Omega, \end{cases}$$

and suppose that $f \in C^1(\mathbb{R})$, $f(0) = 0$ and $f'(0) \neq \lambda_k$. Prove that (P1) has a solution for all $\varepsilon \in \mathbb{R}, |\varepsilon|$ small. This solution is unique in $C^{2,\alpha}(\Omega)$.

8. Prove that the boundary-value problem
$$\left. \begin{array}{c} \Delta u = u^3 \text{ in } \Omega, \\ u = g(x) \text{ on } \partial\Omega, \end{array} \right\}$$
has a unique solution for all $g \in C^{0,\alpha}(\partial\Omega)$.

9. Let $F(u) = \Delta u + \lambda f(u), \lambda \in \mathbb{R}$. Suppose that $f \in C^2(\mathbb{R})$ is such that $f'(0) = 1$ and $uf''(u) < 0$ for all $u \neq 0$. Setting $X = C^{2,\alpha}(\Omega) \cap C_0(\Omega), Y = C^{0,\alpha}(\Omega)$, prove that
(i)   for all $\lambda < \lambda_1$, $F : X \to Y$ is locally invertible on X,
(ii) if, in addition, $f(0) = 0$, then for $\lambda = \lambda_1$ for the singular set of F, Σ_0, one has $\Sigma_0 = \{0\}$.

10. Consider the boundary-value problem

(P2)
$$\begin{cases} \Delta u + \lambda f(u) = h(x) \text{ in } \Omega \\ u = 0 \text{ on } \partial\Omega, \end{cases}$$

where $f \in C^2(\mathbb{R})$ is bounded and such that $f(0) = 0, f'(0) = 1$ and $uf''(u) < 0$ for all $u \neq 0$. Show that (P2) has a unique solution for all $h \in C^{0,\alpha}(\Omega)$.

Hint Use Problems 7 and 9.

11. Extend the preceding results to the case in which f satisfies $f(0) = 0, f'(0) = 1, uf''(u) < 0$ for all $u \neq 0$, but is possibly unbounded. Deduce that for all $h \in C^{0,\alpha}(\Omega)$ the boundary-value problem
$$\left. \begin{array}{c} \Delta u + \lambda u - u^3 = h(x) \text{ in } \Omega, \\ u = 0 \text{ on } \partial\Omega, \end{array} \right\}$$
has a unique solution provided $\lambda \leq \lambda_1$.

12. Consider the boundary value-problem

(P3)
$$\begin{cases} \Delta u + \lambda_k u + b(u) = h(x) \text{ in } \Omega, \\ u = 0 \text{ on } \partial\Omega. \end{cases}$$

Suppose that $b \in C^1(\mathbb{R})$ is bounded and such that $\lambda_{k-1} < c_1 \leq \lambda_k + b'(s) \leq c_2 \leq \lambda_{k+1}$. Let

$$\beta^+ = \lim_{s \to +\infty} \sup b(s), \ \beta^- = \lim_{s \to -\infty} \sup b(s),$$

$$\alpha^+ = \lim_{s \to +\infty} \inf b(s), \ \alpha^- = \lim_{s \to -\infty} \inf b(s)$$

and set

$$A^+ = \alpha^+ \int_{\Omega^+} \phi_k + \beta^- \int_{\Omega^-} \phi_k,$$

$$A^- = \alpha^- \int_{\Omega^+} \phi_k + \beta^+ \int_{\Omega^-} \phi_k,$$

$$B^+ = \beta^+ \int_{\Omega^+} \phi_k + \alpha^- \int_{\Omega^-} \phi_k,$$

$$B^- = \beta^- \int_{\Omega^+} \phi_k + \alpha^+ \int_{\Omega^-} \phi_k$$

Prove that (P3) has a solution provided

$$\min(A^-, A^+) < \int_\Omega h\phi_k < \max(B^-, B^+).$$

13. Consider the boundary-value problem

(P4)
$$\begin{cases} \Delta u + \lambda_1 u + b(u) = a\phi_1 \text{ in } \Omega, \\ u = 0 \text{ on } \partial\Omega, \end{cases}$$

with b satisfying
(i) $b \in C^1(\mathbb{R})$, and $\lambda_1 + b'(s) < \lambda_2$,
(ii) $b^\pm = 0$,
(iii) $b(s) > 0$ for all $s \in \mathbb{R}$.

Prove that there exists $\alpha > 0$ such that (P4) has a solution if and only if $0 < a \leq \alpha$; and, moreover, that if $0 < a < \alpha$ then (P3) has at least two solutions.

14. Consider (P4) with b satisfying (i)–(ii) and
(iv) $sb(s) > 0$ for all $s \neq 0$.
   Prove that there exist $\alpha' < 0 < \alpha''$ such that (P4) has a solution if and only if $\alpha' \leq a \leq \alpha''$; and, moreover, that if $\alpha' < a < \alpha''$ and $a \neq 0$ then (P4) has at least two solutions.

15. Let h be a 2π-periodic contionuous function and consider the problem

(P5) $y'' + f(y) = h(t),\ y(0) - y(2\pi) = y'(0) - y'(2\pi) = 0.$

Suppose that $f \in C^1(\mathbb{R}),\ f'(y) < 1$ and that $f(y) \to f^{\pm}$ as $y \to \pm\infty$. Prove that (P5) has a solution provided $f^- < (2\pi)^{-1} \int_0^{2\pi} h < f^+$.

16. Consider

(P6) $\begin{cases} \Delta u + \lambda_1 u + f(x, u) = 0 \text{ in } \Omega, \\ u = 0 \text{ on } \partial\Omega, \end{cases}$

with $f \in C^1$ satisfying (i) $\lambda_1 + f'(u) < \lambda_2$; keeping the notation of Section 4.1, let

$$G(t) = \int_{\Omega} f(t\phi_1 + w(t))\phi_1.$$

(a) Prove that if $G(t^*) < 0\ (> 0)$ then $u^* = t^*\phi + w(t^*)$ is a super-solution (sub-solution) of (P6).

(b) Use this to show that (P6) has a solution provided it possesses a sub-solution φ and a super-solution ψ (without requiring that $\varphi \leq \psi$).

Hint If G vanishes at some t_0 then $t_0\phi_1 + w(t_0)$ solves (P6). If, say $G(t) < 0$ for all t, then each $t\phi_1 + w(t)$ is a super-solution of (P6). In particular, for $t > 0$ large enough, $t\phi_1 + w(t) > \varphi$.

(c) Find a counterexample of a b.v.p which has a sub-solution φ and a super-solution ψ but does not possess any solution.

Hint Take a problem like $-\Delta u + \lambda_2 u + h(x) = 0$.

17. Consider the (linear) Volterra operator A from $X = C(0, 1)$ into itself, defined by

$$A : u(t) \to \int_0^t u(s)ds.$$

Show that the spectrum $\sigma(A)$ of A contains only $\lambda = 0$, which is not an eigenvalue and hence $\lambda = 0$ is not a bifurcation point. See Remark 5.1.5 (a).

18. Let A be the operator from $X = L^2(a, b)$ into itself defined by

$$A : u(t) \to \int_a^b k(s, t)u(s)ds,$$

where $\Omega = [a, b] \times [a, b]$ and $k(s, t) \in L^2(\Omega)$, is symmetric and positive-definite.

Show that (i) $\lambda = 0$ is not an eigenvalue of A, (ii) A possesses a sequence λ_n of eigenvalues, with $\lambda_n \to 0$, (iii) hence $\lambda = 0$ is a bifurcation point for $F = \lambda I - A$.

19. Discuss the Examples 5.4.5–6 using Theorem 5.4.2 instead of 5.4.1.

20. Consider the system
$$\left.\begin{array}{l} u'' = \lambda \sin \varphi, \\ \varphi'' = \lambda u \cos \varphi, \end{array}\right\}$$
together with the boundary conditions
$$\left.\begin{array}{l} u'(0) = u(1) = 0, \\ \varphi(0) = \varphi'(1) = 0, \end{array}\right\}$$
describing the equilibria of a rotating beam.
   Prove that the solutions λ_k of
$$1 + \cos\sqrt{|\lambda|} \cdot \cosh\sqrt{|\lambda|} = 0$$
are bifurcation points for the above problem.
   Extend the result to a system of the form
$$\left.\begin{array}{l} u'' = \lambda f(u, \varphi), \\ \varphi'' = \lambda g(u, \varphi). \end{array}\right\}$$

21. Let $\varphi, \psi : \mathbb{R} \to \mathbb{R}$ be smooth. Discuss the bifurcation of periodic solutions for the Liénard equation
$$x'' - \varphi(x)x' + \psi(x) = 0,$$
in dependence on the parameter $\mu = \varphi(0)$.
   (Note that for $\varphi(x) = \mu - 3x^2$ and $\psi(x) = x$ this is nothing but the Van der Pol equation).

22. Given $\phi, \psi : \mathbb{R} \to \mathbb{R}$, smooth and such that $\phi(0) = \psi(0) = 0$, discuss the Hopf bifurcation for the system
$$\left.\begin{array}{l} x' = \phi(y), \\ y' = \psi(x) + \mu y. \end{array}\right\}$$

23. Discuss the Lyapunov Centre Theorem in the case of the Hamiltonian system
$$\left.\begin{array}{l} x' = H_y, \\ y' = -H_x, \end{array}\right\}$$
with $H(x, y) = U(x) + V(y)$.

Bibliography

[Ama]   Amann, H., Fixed point equations and nonlinear eigenvalue problems in ordered Banach spaces, *SIAM Review* **18**(1976), 620-709.

[AAM]   Amann, H., Ambrosetti, A. & Mancini, G., Elliptic equations with noninvertible Fredholm linear part and bounded nonlinearities, *Math. Zeit.* **158** (1978), 179-94.

[AmaH]   Amann, H. & Hess, P., A multiplicity result for a class of elliptic boundary value problems, *Proc. Royal Soc. Edinburgh* **84A**(1979), 145-51.

[AM1]   Ambrosetti, A. & Mancini, G., Existence and multiplicity results for nonlinear elliptic problems with linear part at resonance. The case of the simple eigenvalue, *Jour. Diff. Equat.* **28** (1978), 220-45.

[AM2]   Ambrosetti, A. & Mancini, G., Theorems of existence and multiplicity for nonlinear elliptic problems with noninvertible linear part, *Annali Scuola Norm. Sup. Pisa, Serie IV* **5** (1978), 15-28.

[AP]   Ambrosetti, A. & Prodi, G., On the inversion of some differentiable mappings with singularities between Banach spaces, *Ann. Mat. Pura Appl.* **93** (1973), 231-47.

[AmiT]   Amick, C.J. & Turner, R.E.L., A global branch of steady vortex rings, *Jour. Reine Angew. Math.* **384** (1988), 1-23.

[Be]   Berger, M.S., *Nonlinearity and Functional Analysis*, Academic Press, New York, 1977.

[BP]   Berger, M.S. & Podolak, E., On the solutions of a nonlinear Dirichlet problem, *Indiana Univ. Math. Jour.* **24** (1975), 837-46.

[Bö]   Böhme, R., Die Lösung der Verzweigungsgleichungen für nichtlineare Eigenwertprobleme, *Math. Zeit.* **127** (1972), 105-26.

[Br]   Brezis, H., *Analyse fonctionelle, théorie et applications,* Masson Ed., Paris, 1983.

[Ca]   Caccioppoli, R., Un principio di inversione per le corrispondenze funzionali e sue applicazioni alle equazioni alle derivate parziali, *Atti. Acc. Naz. Lincei* **16** (1932), 392-400.

[CH]   Courant, R. & Hilbert, D., *Methods of Mathematical Physics* , Interscience, New York, 1962.

[ChR]   Chow, S.N. & Hale, J.K., *Methods of Bifurcation Theory,* Springer-Verlag, New York, 1982.

[CrR]   Crandall, M. & Rabinowitz, P.H., Nonlinear Sturm-Liouville eigenvalue problems and topological degree, *Jour. Math. Mechanics* **19** (1970), 1083-1102.

[DST]   DeSimon, L. & Torelli, G., Soluzioni periodiche di equazioni alle derivate parziali di tipo iperbolico nonlineari, *Rend. Sem. Mat. Univ. Padova* **40** (1968), 380-401.

[D1]   Dieudonné, J., *Éléments d'analyse,* Gautheir-Villars, Paris, 1969.

[D2]   Dieudonné, J., Sur le polygone de Newton, *Arkiv der math.* **2** (1950), 49-55.

[Fi]   Field, M.J., *Differential Calculus and its Applications,* Van Nostrand Reinhold, New York, 1976.

[Fu]   Fučik, S., *Solvability of Nonlinear Equations and Boundary Value Problems,* Reidel, Dordrecht, 1980.

[GT]   Gilbarg, D. & Trudinger, N., *Elliptic Partial Differential Equations of Second Order,* Springer-Verlag, New York, 1977.

[GS]   Golubitski, M. & Schaeffer, D., A theory of imperfect bifurcation, *Comm. Pure Appl. Math.* **32** (1974), 21-98.

[Ho]   Hopf, E., Abzweigung einer periodischen Lösung von einer stationären Lösung eines Differentialsystemes, *Ber. Math. Phys. Sachsische Akademie der Wissenschaften Leipzig* **94** (1942), 1-22.

[KW]   Kazdan, J.L. & Warner, F.W., Remarks on some quasilinear elliptic equations, *Comm. Pure Appl. Math* **28** (1975), 567-97.

[Ko]   Kolodner, I.I., Heavy rotating string, a nonlinear problem, *Comm. Pure Appl. Math.* **8** (1955), 395-408.

[Kr1]   Krasnoselski, M.A., *Topological Methods in the Theory of Nonlinear Integral Equations*, Pergamon, Oxford, 1965.

[Kr2]   Krasnoselski, M.A., *Positive Solutions of Operator Equations*, Noordhoff, 1964.

[KFS]   Kufner, A., John, O. & Fučik, S., *Function spaces*, Academia, Prague, 1977.

[LL]   Landesman, E. & Lazer, A.C., Nonlinear perturbations of linear eigenvalue problems at resonance, *Jour. Math. Mechanics* **19** (1970), 609-23.

[LC]   Levi-Civita, T., Détermination rigoureuse des ondes d'ampleur finie, *Math. Annalen* **93** (1925), 264-314.

[Le]   Levy, P., Sur les fonctions de lignes implicites, *Bull. Soc. Math. de France* **48** (1920).

[Ly1]   Lyapunov, A.M., Sur les figures d'équilibre peu différents des ellipsoïdes d'une masse liquide homogène douée d'un mouvement de rotation, *Zap. Akad. Nauk St. Petersburg* (1906), 1-225.

[Ly2]   Lyapunov, A.M., Problème général de la stabilité du mouvement, *Ann. Fac. Sci. Toulouse* **2** (1907), 203-474.

[Mar]   Marino, A., La biforcazione nel caso variazionale, *Conf. Sem. Mat. Univ. Bari* **132** (1973).

[MP]   Marino, A. & Prodi, G., La teoria di Morse per gli spazi di Hilbert, *Rend. Sem. Mat. Univ. Padova* **41** (1968), 43-68.

[MM]   Marsden, J. & McCracken, M., *The Hopf Bifurcation and its Applications*, Springer-Verlag, New York, 1976.

[Maw]   Mawhin, J., Topological degree methods in nonlinear boundary value problems, *CBMS Regional Conference Series Math. #40*, A.M.S., Providence, R.I., 1977.

[MW]   Mawhin, J. & Willem, M., *Critical Point Theory and Hamiltonian Systems*, Springer-Verlag, New York, 1989.

[McS]   McKean, H.P. & Scovel, J.C., Geometry of some simple non-linear differential operators, *Annali Scuola Norm. Sup. Pisa, Serie IV* **13** (1986), 299-346.

[Mi]   Minty, G.J., Monotone (nonlinear) operators in Hilbert spaces, *Duke Math. Jour.* **29** (1962), 341-6.

[Mo1]   Moser, J., Lectures on Hamiltonian Systems, *Mem. A.M.S. 68* (1968).

[Mo2]   Moser, J., Periodic orbits near an equilibrium and a theorem of A. Weinstein, *Comm. Pure Appl. Math.* **29** (1976), 727-47.

[Ne]   Nekrasov, A.I., Waves of stationary type (in Russian), *Izv. Ivanovo. Voznesensk pol. Inst.* **6** (1922), 155-71.

[Ni]   Nirenberg, L., *Topics in Nonlinear Functional Analysis,* New York Univ. Lecture Notes, 1974.

[P]   Perron, G., Über Stabilität und asymptotische Verhalten der Integrales von Differentialgleichungsystemen, *Math. Zeit.* **29** (1929), 748-66.

[Pr]   Prodi, G., Problemi di diramazione per equazioni funzionali, *Boll. U.M.I.* **22** (1967), 413-33.

[PW]   Protter, M.H. & Weinberger, H.F. *Maximum Principles in Differential Equations,* Prentice-Hall, Englewood Cliffs, N.J., 1967.

[R1]   Rabinowitz, P.H., Periodic solutions of nonlinear hyperbolic partial differential equations, *Comm. Pure Appl. Math.* **20** (1967), 145-205.

[R2]   Rabinowitz, P.H., Some global results for nonlinear eigenvalue problems, *Jour. Funct. Anal.* **7** (1971), 487-513.

[R3]   Rabinowitz, P.H., Existence and nonuniqueness of rectangular solutions of the Bénard problem, *Arch. Rat. Mech. Anal.* **29** (1968), 32-57.

[Sa]   Sansone, G., *Equazioni differenziali nel campo reale,* Zanichelli, Bologna, 1965.

[Sche]   Schechter, M., A nonlinear elliptic boundary value problem, *Ann. Scuola Norm. Sup. Pisa* **27** (1973), 707-16.

[Schm]   Schmidt, E., Zur Theorie der linearen und nichtlinearen Integralgleichungen, 3 Teil, *Math. Annalen* **65** (1908), 370-99.

[**Schw**] Schwartz, J.T., *Nonlinear Functional Analysis,* Gordon & Breach, New York, 1953.

[**SiM**] Siegel, C.L. & Moser, J., *Lectures on Celestial Mechanics,* Springer-Verlag, 1971.

[**T**] Turner, R.E.L., Internal waves in fluids with rapidly varying density, *Ann. Scuola Norm. Sup. Pisa, Serie IV, Vol. VIII*-4 (1981), 513-73.

[**Va**] Vainberg, M.M., *Variational Methods for the Study of Nonlinear Operators,* Holden-Day, San Francisco, 1964.

[**VT**] Vainberg, M.M. & Trenogin, V.A., The methods of Lyupanov and Schmidt in the theory of nonlinear equations and their further development, *Uspehi Mat. Mauk* **17** (1962), 13-75; English translation in *Russian Math. Surveys.*

[**Ve**] Velte, W., Stabilität und Verzweigung stationären Lösungen der Navier-Stokesschen Gleichungen beim Taylorproblem, *Arch. Rat. Mech. Anal.* **22** (1966), 1-14.

[**W**] Weinstein, A., Normal modes for non-linear Hamiltonian systems, *Inv. Math.* **20** (1973), 47-57.

[**Y**] Yosida, K., *Functional Analysis,* Springer-Verlag, New York, 1974.

Index